T0229805

Secure and Smart Internet of Things (IoT)
Using Blockchain and AI

RIVER PUBLISHERS SERIES IN INFORMATION SCIENCE AND TECHNOLOGY

Series Editors:

K. C. CHEN
National Taiwan University
Taipei, Taiwan
and
University of South Florida, USA

SANDEEP SHUKLA
Virginia Tech
USA
and
Indian Institute of Technology Kanpur, India

Indexing: All books published in this series are submitted to the Web of Science Book Citation Index (BkCI), to CrossRef and to Google Scholar.

The "River Publishers Series in Information Science and Technology" covers research which ushers the 21st Century into an Internet and multimedia era. Multimedia means the theory and application of filtering, coding, estimating, analyzing, detecting and recognizing, synthesizing, classifying, recording, and reproducing signals by digital and/or analog devices or techniques, while the scope of "signal" includes audio, video, speech, image, musical, multimedia, data/content, geophysical, sonar/radar, bio/medical, sensation, etc. Networking suggests transportation of such multimedia contents among nodes in communication and/or computer networks, to facilitate the ultimate Internet.

Theory, technologies, protocols and standards, applications/services, practice and implementation of wired/wireless networking are all within the scope of this series. Based on network and communication science, we further extend the scope for 21st Century life through the knowledge in robotics, machine learning, embedded systems, cognitive science, pattern recognition, quantum/biological/molecular computation and information processing, biology, ecology, social science and economics, user behaviors and interface, and applications to health and society advance.

Books published in the series include research monographs, edited volumes, handbooks and textbooks. The books provide professionals, researchers, educators, and advanced students in the field with an invaluable insight into the latest research and developments.

Topics covered in the series include, but are by no means restricted to the following:

- Communication/Computer Networking Technologies and Applications
- Queuing Theory
- Optimization
- Operation Research
- Stochastic Processes
- Information Theory
- Multimedia/Speech/Video Processing
- Computation and Information Processing
- Machine Intelligence
- Cognitive Science and Brian Science
- Embedded Systems
- Computer Architectures
- Reconfigurable Computing
- Cyber Security

For a list of other books in this series, visit www.riverpublishers.com

Secure and Smart
Internet of Things (IoT)
Using Blockchain and AI

Ahmed Banafa

Professor of Engineering at San Jose State University

USA

and

Instructor of Continuing Studies at Stanford University

USA

River Publishers

Published, sold and distributed by:
River Publishers
Alsbjergvej 10
9260 Gistrup
Denmark

River Publishers
Lange Geer 44
2611 PW Delft
The Netherlands

Tel.: +45369953197
www.riverpublishers.com

ISBN: 978-87-7022-030-9 (Hardback)
      978-87-7022-029-3 (Ebook)

©2018 River Publishers

Contents

Foreword   **xi**

Preface   **xv**

List of Figures   **xvii**

List of Abbreviations   **xix**

PART I: What Is IoT?

1   Internet of Things: The Third Wave?   **3**
    1.1   Advantages and Disadvantages of IoT   5

2   The Industrial Internet of Things (IIoT): Challenges,
Requirements and Benefits   **7**
    2.1   IIoT, IoT and M2M   9
        2.1.1   IIoT Challenges   9
        2.1.2   IIoT Requirements   9
        2.1.3   IIoT Benefits   10
    2.2   The Future of the Industrial Internet of Things   11

3   Internet of Things: Security, Privacy and Safety   **13**
    3.1   IoT's Threats   15
    3.2   Threats Are Real   16
    3.3   What Can We Do   17

4   Internet of Things: More than Smart "Things"   **19**
    4.1   Industrial Internet   21
    4.2   The IoT Value Chain   23
    4.3   Internet of Things Predictions   23
    4.4   Challenges Facing IoT   24

5   Internet of Things: Myths and Facts   **25**
   5.1   IoT and Sensors   26
   5.2   IoT and Mobile Data   27
   5.3   IoT and the Volume of Data   28
   5.4   IoT and Datacenters   28
   5.5   IoT Is a Future Technology   28
   5.6   IoT and Current Interoperability Standards   29
   5.7   IoT and Privacy & Security   30
   5.8   IoT and Limited Vendors   30
   5.9   Conclusion   31

PART II: IoT Implementation & Standardization Challenges

6   Three Major Challenges Facing IoT   **35**
   6.1   Technology   36
   6.2   Technological Challenges   37
   6.3   Business   40
   6.4   Society   41
   6.5   Privacy   42
   6.6   Regulatory Standards   43

7   IoT Implementation and Challenges   **45**
   7.1   Components of IoT Implementation   46
      7.1.1   Sensors   46
      7.1.2   Networks   47
      7.1.3   Standards   48
      7.1.4   Intelligent Analysis   49
      7.1.5   Intelligent Actions   51

8   IoT Standardization and Implementation Challenges   **53**
   8.1   IoT Standardization   54
   8.2   IoT Implementation   55
      8.2.1   Sensors   55
      8.2.2   Networks   56
      8.2.3   Standards   57
      8.2.4   Intelligent Analysis   57
      8.2.5   Intelligent Actions   58
   8.3   The Road Ahead   58

9 Challenges Facing IoT Analytics Success   **59**
     9.1   Data Structures 61
     9.2   Combining Multiple Data Formats 61
     9.3   The Need to Balance Scale and Speed 61
     9.4   IoT Analytics at the Edge   62
     9.5   IoT Analytics and AI 63

PART III: Securing IoT

10 How to Secure the Internet of Things   **69**
     10.1   Challenges to Secure IoT Deployments   72
     10.2   Dealing with the Challenges and Threats   74
     10.3   The Optimum Platform 74
     10.4   Last Word 75

11 Using Blockchain to Secure IoT   **77**
     11.1   Challenges to Secure IoT Deployments   77
     11.2   Dealing with the Challenges and Threats   78
     11.3   The Optimum Platform 78
     11.4   Decentralizing IoT Networks 79
     11.5   The Blockchain Approach   80
          11.5.1   What Is Blockchain? 81
          11.5.2   What Are Some Advantages of Blockchain? 81
          11.5.3   How Does It Work? 82
     11.6   The Blockchain and IoT   83

12 IoT and Blockchain: Challenges and Risks   **85**
     12.1   The Blockchain Model 86
     12.2   Principles of Blockchain Technology   87
     12.3   Public vs. Private Blockchain   88
     12.4   Challenges of Blockchain in IoT 88
     12.5   Risks of Using Blockchain in IoT   89
     12.6   The Optimum Secure IoT Model 91

13 DDoS Attack: A Wake-Up Call for IoT   **93**
     13.1   Security Is Not the Only Problem   95
          13.1.1   Security Concerns 95
          13.1.2   Privacy Issues 96
          13.1.3   Inter-Operatability Standard Issues 96

13.1.4 Legal Regulatory and Rights Issues 97
13.1.5 Emerging Economy and Development Issues 97
13.2 How to Prevent Future Attacks? 97

PART IV: AI, Fog Computing and IoT

14 Why IoT Needs Fog Computing 101
14.1 The Challenge 101
14.2 The Solution 102
14.3 Benefits of Using Fog Computing 103
14.4 Real-Life Example 103
14.5 The Dynamics of Fog Computing 104
14.6 Fog Computing and Smart Gateways 105
14.7 Conclusion 105

15 AI Is the Catalyst of IoT 107
15.1 Examples of IoT Data 109
15.2 AI and IoT Data Analysis 110
15.3 AI in IoT Applications 110
15.4 Challenges Facing AI in IoT 111
15.5 Conclusion 112

PART V: The Future of IoT

16 IoT, AI and Blockchain: Catalysts for Digital Transformation 115
16.1 Digital Transformation 116
16.2 Internet of Things (IoT) 117
16.3 Digital Transformation, Blockchain and AI 118
16.4 Conclusion 118

17 Future Trends of IoT 121
17.1 Trend 1 – Lack of Standardization Will Continue 122
17.2 Trend 2 – More Connectivity and More Devices 122
17.3 Trend 3 – "New Hope" for Security: IoT & Blockchain
Convergence 123
17.4 Trend 4 – IoT Investments Will Continue 124
17.5 Trend 5 – Fog Computing Will Be More Visible 124
17.6 Trend 6 – AI & IoT Will Work Closely 125

17.7 Trend 7 – New IoT-as-a-Service (IoT-a-a-S) Business
Models 125
17.8 Trend 8 – The Need for Skills in IoT's Big Data Analysis and
AI Will Increase 126

PART VI: Inside Look at Blockchain

18 Myths about Blockchain Technology   **129**
18.1 Myth 1: The Blockchain Is a Magical Database
in the Cloud 132
18.2 Myth 2: Blockchain Is Going to Change the World 132
18.3 Myth 3: Blockchain Is Free 133
18.4 Myth 4: There Is Only One Blockchain 133
18.5 Myth 5: The Blockchain Can be Used for Anything and
Everything 133
18.6 Myth 6: The Blockchain Can be the Backbone of a Global
Economy 134
18.7 Myth 7: The Blockchain Ledger Is Locked and Irrevocable . 134
18.8 Myth 8: Blockchain Records Can Never be Hacked
or Altered 134
18.9 Myth 9: Blockchain Can Only be Used in the Financial
Sector 135
18.10 Myth 10: Blockchain Is Bitcoin 135
18.11 Myth 11: Blockchain Is Designed for Business Interactions
Only 136
18.12 Myth 12: Smart Contracts Have the Same Legal Value as
Regular Contracts 136

19 Cybersecurity & Blockchain   **137**
19.1 Implementing Blockchain in Cybersecurity 138
19.2 Advantages of Using Blockchain in Cybersecurity 139
19.2.1 Decentralization 139
19.2.2 Tracking and Tracing 140
19.2.3 Confidentiality 140
19.2.4 Fraud Security 140
19.2.5 Sustainability 141
19.2.6 Integrity 141
19.2.7 Resilience 141
19.2.8 Data Quality 141

19.2.9 Smart Contracts 141
19.2.10 Availability 141
19.2.11 Increase Customer Trust 142
19.3 Disadvantages of Using Blockchain in Cybersecurity 142
19.3.1 Irreversibility 142
19.3.2 Storage Limits 142
19.3.3 Risk of Cyberattacks 142
19.3.4 Adaptability Challenges 142
19.3.5 High Operation Costs 143
19.3.6 Blockchain Literacy 143
19.4 Conclusion 143

20 Future Trends of Blockchain   **145**
20.1 Blockchain Tracks 146
20.2 Blockchain Technology Future Trends 147
20.3 Blockchain Skills in Demand 147
20.4 Blockchain and Enterprise Applications 148
20.5 Blockchain and IoT Security 149
20.6 Blockchain and Zero-Trust Model 150
20.7 Final Words 150

References   **151**

Index   **159**

About the Author   **163**

Foreword

The Internet of Things (IoT) connects everything on our bodies, in our homes, cities, and factories to the Internet to create a new reality of a connected world. Imagine a person driving an autonomous vehicle driving though connected traffic lights wearing a connected pacemaker. Every connected thing comes with security risk with the potential for hackers to target the person. Even security updates of AI inference models on these devices can be hacked when the device makers send an OTA or Over the Air update.

Blockchain gives you a distributed ledger in the cloud that is secure from tampering because it is cryptographically secured with transparent access to everyone equally.

Blockchain is used with the Internet of Thing (IoT) systems to track every access point of the device thereby ensuring security. Blockchain offers additional transparency to a supply chain or logistics process by offering new insights to where the products originated and how it traveled to the endpoint. Imagine knowing which farm produced the coffee you are drinking and every step it traveled to reach your Barista. Blockchain also helps secure complex systems such as mines, hospitals and critical infrastructures such as water and energy systems.

I have watched Prof. Banafa's passion for IoT over the past years as he researched IoT markets with special focus on Industrial IoT with his articles, talks, and teaching.

Read this book to understand IoT solutions, as Prof. Banafa demystifies the grey areas of Edge, Fog and Cloud and drills down to provide clarity to the integration of new technologies such as Blockchain and Artificial Intelligence on how they fit to make IoT secure. Treasure this book as you will be re-reading and coming back to it several times as you build the connected world of the Internet of Things.

Sudha Jamthe
CEO, Author IoT Disruptions &
Stanford Continuing Studies IoT and AV Business Instructor

We live in exciting times where technology emergence is making us reinvent the world around us. There is no single technology that can help this transformation, but the confluence of many emerging technologies such as IoT, AI, Blockchain, AR/VR, Edge Computing will enable the current generation to visualize and discover new ways of doing things, that were earlier not possible. And this transformation will cut across every aspect of our life: Consumer to Enterprise, Community to Government, Homes to Hospitals, Art to Law, Engineering to Medicine, Infants to Elderly, Poor to Rich, impacting every form of life and resource on this planet and beyond.

And that bring us face to face with the challenges that it will bring relating to privacy and security, and the associated vulnerabilities that the new technologies bring. The human mind has, however, always fought these challenges and emerged successful, rather than letting these challenges consume us. The Blockchain is an ingenious approach that provides secure and trustless decentralized framework that holds the promise to provide an impregnable and transparent transaction system.

"Secure and Smart Internet of Things (IoT): Using Blockchain and AI" by Prof. Banafa is one such book that has covered various types of emerging technologies, their convergence, and the supporting technologies like Mist and Fog Computing, that have emerged to take AI from the cloud to the edge. I have known Prof. Banafa for many years and have a deep appreciation for his ability to mix academics with industry use-cases to make them relevant for applications in various domains. Apart from being a "#1 Voice to Follow in Tech" on LinkedIn, his prophecies of technology evolution have been very accurate. In this book, he shares his valuable technology insights that will be very useful to aspiring innovators to set them thinking on the right path.

My advice to readers is to relate the technologies discussed here, in context of their own ideas and visions, to discover how they can play an instrumental role in building the emerging new world that will bring peace, joy, and happiness to all.

Sudhir Kadam
Startup Alchemist, Silicon Valley Emerging Tech Expert
Board Member at IIT Startups

In Secure and Smart Internet of Things (IoT) using Blockchain and AI, Ahmed, a seasoned and revered stalwart on IoT and cybersecurity chronicles the vanguard of what science fictionists once called Skynet: the set of technologies converging at warp speed to create a massive intelligent machine or device network. The revolution of "Internet of Things" is amassing to some 28 billion devices by 2020 according to some experts. This ubiquity presents a "Cambrian explosion": an unprecedented economic opportunity and is consequently poised to spark an arms race in cybercrime or cyberwars. It is crucial to grasp the magnitude of the cybersecurity risks and the challenge device manufacturers face. Imagine for a moment the chaos that would ensue when a rogue entity commandeers a system of autonomous vehicles operating in a smart city. Worse yet, imagine a world of interconnected healthcare systems being hijacked and mission critical information for patient treatment is held ransom or is compromised. There is no question that we approach a world where cybercrime and cyberwar is dire. It's no longer science fiction. Thanks to Ahmed who has done most of the homework for us in presenting a cogent methodology to counter this formidable threat. To anyone from (smart)home-owners to business-owners and in between – Secure and Smart Internet of Things using Blockchain and AI is the call to action!

Abood Quraini
Sr. Manager, Hardware Engineering at NVIDIA

Preface

By 2020, experts forecast that up to 28 billion devices will be connected to the Internet with only one third of them being computers, smartphones, smartwatches, and tablets. The remaining two thirds will be other "devices" – sensors, terminals, household appliances, thermostats, televisions, automobiles, production machinery, urban infrastructure and many other "things", which traditionally have not been Internet enabled.

This "Internet of Things" (IoT) represents a remarkable transformation of the way in which our world will soon interact. Much like the World Wide Web connected computers to networks, and the next evolution mobile devices connected people to the Internet and other people, IoT looks poised to interconnect devices, people, environments, virtual objects and machines in ways that only science fiction writers could have imagined. In a nutshell the Internet of Things (IoT) is the convergence of connecting people, things, data and processes; it is transforming our life, business and everything in between. This book *Secure and Smart IoT* explores many aspects of the Internet of Things and explains many of the complicated principles of IoT and the new advancements in IoT including using Fog Computing, AI and Blockchain technology.

This book is divided into 6 parts:

Part 1: What Is IoT?
Part 2: IoT Implementation & Standardization Challenges
Part 3: Securing IoT
Part 4: AI, Fog Computing and IoT
Part 5: The Future of IoT
Part 6: Inside Look at Blockchain

Audience

This book is for everyone who would like to have a good understanding of IoT and its applications and its relationship with business operations including: C-Suite executives, IT managers, marketing & sales people, lawyers, product & project managers, business specialists, students. It's not for programmers who are looking for codes or exercises on the different components of IoT.

Acknowledgment

I would like to thank my mother, my father and my wonderful family for all their support during the stages of writing this book. I love you all.

List of Figures

Figure 2.1	IIoT Challenges.	10
Figure 2.2	IIoT Requirements.	10
Figure 2.3	IIoT Benefits.	11
Figure 4.1	IoT Key Attributes.	22
Figure 6.1	Three Major Challenges Facing IoT.	36
Figure 6.2	Technological Challenges Facing IoT.	36
Figure 6.3	Categories of IoT.	42
Figure 7.1	Components of IoT implementation.	46
Figure 7.2	New trends of sensors.	47
Figure 8.1	Hurdles facing IoT standardization.	54
Figure 8.2	Components of IoT implementations.	56
Figure 9.1	Challenges facing IoT analytics.	60
Figure 9.2	AI and IoT Data Analysis.	64
Figure 10.1	IoT architecture.	70
Figure 11.1	Advantages of Blockchain.	81
Figure 12.1	Challenges of Blockchain in IoT.	89
Figure 12.2	Risks of using Blockchain in IoT.	90
Figure 13.1	How to prevent future attacks.	98
Figure 15.1	Challenges facing AI in IoT.	112
Figure 16.1	Digital Transformation Areas.	116
Figure 17.1	Future Trends of IoT.	121
Figure 18.1	Types of Blockchains.	131
Figure 19.1	Advantages of Using Blockchain in Cybersecurity.	140
Figure 19.2	Disadvantages of Using Blockchain in Cybersecurity.	142
Figure 20.1	Tracks of Blockchain Technology.	146
Figure 20.2	Blockchain Technology Future Trends.	147

List of Abbreviations

AI	Artificial Intelligence
BYOD	Bring Your Own Device
CEP	Complex Event Processing
CSA	Cloud Security Alliance
DDoS	Distributed Denial of Service
DoS	Denial of Service
DX	Digital Transformation
E-a-a-S	Elevator as a Service
ETL	Extract, Transform and Load
FIPP	Fair Information Practice Principles
FTC	Federal Trade Commission
HIPAA	Health Insurance Portability and Accountability Act
HTTP	Hypertext Transfer Protocol
IaaS	Infrastructure as a Service
IDC	International Data Corporation
IEEE	Institute of Electrical and Electronics Engineering
IIoT	Industrial Internet of Things
IoT	Internet of Things
IP	Internet Protocol
IPv6	Internet Protocol Version 6
ISA	International Society of Automation
ISOC	Information Security Operations Center
IT	Information Technology
L-a-a-S	Lighting as a Service
LAN	Local Area Network
Li-Fi	Light Fidelity
LTE	Long-Term Evolution
M2H	Machine to Human
M2M	Machin to Machine
MAN	Merto Area Network
OEM	Original Equipment Manufacturer

OWASP	Open Web Application Security Project
PaaS	Platform as a Service
PSS	Privacy, Security, Safety
R-a-a-S	Rail as a Service
SaaS	Software as a Service
UI	User Interface
UX	User Experience
VUI	Voice User Interface
W3C	World Wide Web Consortium
WAN	Wide Area Network
Wi-Fi	Wireless Fidelity
ZTM	Zero Trust Model

PART I

What Is IoT?

1

Internet of Things: The Third Wave?

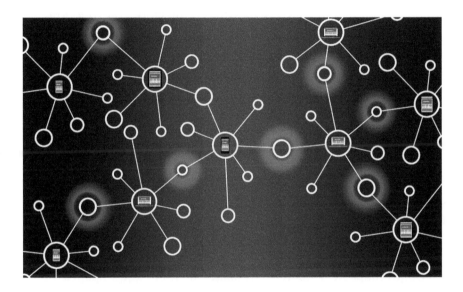

The Internet of Things (IoT) is the network of physical objects accessed through the Internet. These objects contain embedded technology to interact with internal states or the external environment. In other words, when objects can sense and communicate, it changes how and where decisions are made, and who makes them, for example, Nest thermostats.

The Internet of Things (IoT) emerged as the third wave in the development of the Internet. The Internet wave of the 1990s connected 1 billion users, while the mobile wave of the 2000s connected another

2 billion. The IoT has the potential to connect 10 times as many 20 billion "things" to the Internet by 2020, ranging from bracelets to cars [17]. Breakthroughs in the cost of sensors, processing power and bandwidth to connect devices are enabling ubiquitous connections at present. Smart products like smart watches and thermostats (Nest) are already gaining traction as stated in the Goldman Sachs Global Investment Research report.

IoT has key attributes that distinguish it from the "regular" Internet, as captured by Goldman Sachs's S-E-N-S-E framework: *Sensing, Efficient, Networked, Specialized, Everywhere* [1]. These attributes may change the direction of technology development and adoption, with significant implications for Tech companies – much like the transition from the fixed to the mobile Internet shifted the center of gravity from Intel to Qualcomm or from Dell to Apple.

A number of significant technology changes have come together to enable the rise of the IoT. These include the following [1]:

- *Cheap sensors* – Sensor prices have dropped to an average 60 cents from $1.30 in the past 10 years.
- *Cheap bandwidth* – The cost of bandwidth has also declined precipitously, by a factor of nearly 40 times over the past 10 years.
- *Cheap processing* – Similarly, processing costs have declined by nearly 60 times over the past 10 years, enabling more devices to be not just connected, but smart enough to know what to do with all the new data they are generating or receiving.
- *Smartphones* – Smartphones are now becoming the personal gateway to the IoT, serving as a remote control or hub for the connected home, connected car or the health and fitness devices that consumers have increasingly started to wear.
- *Ubiquitous wireless coverage* – With Wi-Fi coverage now ubiquitous, wireless connectivity is available for free or at a very low cost, given Wi-Fi utilizes unlicensed spectrum and thus does not require monthly access fees to a carrier.

- *Big Data* – As the IoT will, by definition, generate voluminous amounts of unstructured data, the availability of Big Data analytics is a key enabler.
- *IPv6* – Most networking equipment now support IPv6, the newest version of the Internet Protocol (IP) standard that is intended to replace IPv4. IPv4 supports 32-bit addresses, which translates to about 4.3 billion addresses – a number that has become largely exhausted by all the connected devices globally. In contrast, IPv6 can support 128-bit addresses, translating to approximately 3.4 × 1038 addresses – an almost limitless number that can amply handle all conceivable IoT devices.

1.1 Advantages and Disadvantages of IoT

Many smart devices like laptops, smart phones and tablets communicate with each other through the use of Wi-Fi Internet technology. Transfer these technological capabilities into ordinary household gadgets like refrigerators, washing machines, microwave ovens, thermostat, door locks among others; equip these with their own computer chips, software and get access to the Internet, and a "smart home" now comes to life.

The Internet of Things can only work if these gadgets and devices start interacting with each other through a networked system. The AllJoyn Open Source Project [2], a nonprofit organization devoted to the adoption of the Internet of Things, facilitates to ensure that companies like Cisco, Sharp and Panasonic are manufacturing products compatible with a networked system and to ensure that these products can interact with each other.

The advantages of these highly networked and connected devices mean productive and enhanced quality of lives for people. For example, health monitoring can be rather easy with connected RX bottles and medicine cabinets. Doctors supervising patients can monitor their medicine intake as well as measure blood pressure and sugar levels and alert them when something goes wrong to their patients online.

In the aspect of energy conservation, household appliances can suggest optimal setting based on the user's energy consumption like turning the ideal temperature just before the owner arrives home as well as turning on and off the lights whenever the owner is out on vacation just to create the impression that somebody is still left inside the house to prevent burglars from attempting to enter.

Smart refrigerators, on the other hand, can suggest food supplies that are low on inventory and needs immediate replenishment. The suggestions are based on the user's historical purchasing behavior and trends. Wearable technology are also part of the Internet of Things, where these devices can monitor sleeping patterns, workout measurements, sugar levels, blood pressure and connecting these data to the user's social media accounts for tracking purposes.

The most important disadvantage of the Internet of Things is with regard to the privacy and security issue. Smart home devices have the ability to devour a lot of data and information about a user. These data can include personal schedules, shopping habits, medicine intake schedule and even location of the user at any given time. If these data fall are misused, great harm and damage can be done to people.

The other disadvantage is the fact that most devices are not yet ready to communicate with another brand of devices. Specific products can only be networked with their fellow products under the same brand name. It is good that The AllJoyn Open Source Project [2], is ensuring connectivity, but the reality of a "universal remote control" for all these devices and products is still in its infancy.

2

The Industrial Internet of Things (IIoT): Challenges, Requirements and Benefits

The idea of a smarter world, where systems with sensors and local processing are connected to share information, is taking hold in every single industry. These systems will be connected on a global scale with users and each other to help them make more informed decisions. Many labels have been given to this overarching idea, but the most ubiquitous is the Internet of Things (IoT). The IoT includes everything from smart homes, mobile fitness devices and connected toys to the

Industrial Internet of Things (IIoT) with smart agriculture, smart cities, smart factories and the smart grid.

The Industrial Internet of Things (IIoT) is a network of physical objects, systems, platforms and applications that contain embedded technology to communicate and share intelligence with each other, the external environment and with people. The adoption of the IIoT is being enabled by the improved availability and affordability of sensors, processors and other technologies that have helped facilitate capture of and access to real-time information.

The IIoT can be characterized as a vast number of connected industrial systems that are communicating and coordinating their data analytics and actions to improve industrial performance and benefit society as a whole. Industrial systems that interface the digital world to the physical world through sensors and actuators that solve complex control problems are commonly known as cyber-physical systems. These systems are being combined with Big Data solutions to gain deeper insight through data and analytics.

Imagine industrial systems that can adjust to their own environments or even their own health. Instead of running to failure, machines schedule their own maintenance or, better yet, adjust their control algorithms dynamically to compensate for a worn part and then communicate those data to other machines and the people who rely on those machines. By making machines smarter through local processing and communication, the IIoT could solve problems in ways that were previously inconceivable. But, as the saying goes, "If it was easy, everyone would be doing it." As innovation grows, so does the complexity, which makes the IIoT face a significant challenge that no company alone can meet.

At its root, the IIoT is a vast number of connected industrial systems that communicate and coordinate their data analytics and actions to improve performance and efficiency and reduce or eliminate downtime. A classic example is industrial equipment on a factory floor that can detect minute changes in its operations, determine the probability of a component failure and then schedule maintenance of that component before its failure can cause unplanned downtime that could cost millions of dollars.

The possibilities in the industrial space are nearly limitless: smarter and more efficient factories, greener energy generation, self-regulating buildings that optimize energy consumption, cities that adjust and can adjust traffic patterns to respond to congestion and more. But, of course, implementation will be a challenge [3].

2.1 IIoT, IoT and M2M

The main difference between IoT and IIoT is that where consumer IoT often focuses on convenience for individual consumers, while Industrial IoT is strongly focused on improving the efficiency, safety and productivity of operations with a focus on return on investment [4]. M2M is a subset of IIoT, which tends to focus very specifically on machine-to-machine communications, where IoT expands that to include machine-to-object/people/infrastructure communications. The IIoT is about making machines more efficient and easier to monitor.

2.1.1 IIoT Challenges (Figure 2.1)

- Precision
- Adaptability and scalability
- Security
- Maintenance and updates
- Flexibility

2.1.2 IIoT Requirements (Figure 2.2)

- Cloud computing
- Access (anywhere, anytime)
- Security
- Big Data analytics
- UX (user experience)
- Assets management
- Smart machines

Figure 2.1   IIoT Challenges.

Figure 2.2   IIoT Requirements.

2.1.3 IIoT Benefits (Figure 2.3)

- Vastly improved operational efficiency (e.g., improved uptime, asset utilization) through predictive maintenance and remote management
- The emergence of an outcome economy, fueled by software-driven services; innovations in hardware; and the increased visibility into products, processes, customers and partners
- New connected ecosystems, coalescing around software platforms that blur traditional industry boundaries

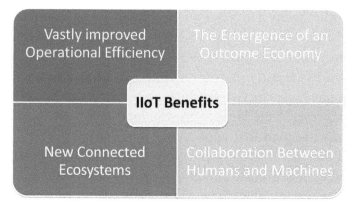

Figure 2.3 IIoT Benefits.

- Collaboration between humans and machines, which will result in unprecedented levels of productivity and more engaging work experiences

2.2 The Future of the Industrial Internet of Things

Accenture estimates that it could add more than $10 trillion to the global economy by 2030 [5]. And that number could be even higher if companies were to take bolder actions and make greater investments in innovation and technology than they are doing today.

The good news is the Industrial Internet of Things is already here, at least among the most forward-thinking companies. The challenge is that most businesses are not ready to take the plunge. According to an Accenture survey of more than 1,400 business leaders, only one-third (36%) claim they fully grasp the implications of the IIoT. Just 7% have developed a comprehensive IIoT strategy with investments to match.

One of the reasons is the as-yet limited ability to leverage machine intelligence to do more than enhance efficiencies on the factory floor and evolve to create entirely new value-added services, business models and revenue streams.

So far, businesses have made progress in applying the Industrial Internet of Things to reduce operational expenses, boost productivity or improve worker safety. Drones, for example, are being used to monitor remote pipelines, and intelligent drilling equipment can improve productivity in mines. Although these applications are valuable, they are reminiscent of the early days of the Internet, when the new technology was limited primarily to speeding up work processes. As with the Internet, however, there is more growth, innovation and value that can be derived with smart IIoT applications.

Imagine a building management company charging fees based on the energy savings it delivers to building owners. Or an airline company rewarding its engine supplier for reduced passenger delays resulting from performance data that automatically schedules maintenance and orders spare parts while a plane is still in flight. With IIoT, there will be no more missing planes, information is live and up-to-date about the plane and the need for the black box will diminish. These are the kinds of product–service hybrid models that can provide new value to customers.

This transformation in business will also have dramatic implications for the workforce. Clearly, the Industrial Internet of Things will digitize some jobs that have, until now, resisted automation. But the vast majority of executives we surveyed believe that the IIoT will be a net creator of jobs. Perhaps more importantly, routine tasks will be replaced by more engaging work, as technology allows workers to do more. As the focus shifts from products to customers, knowledge-intensive work will be required to handle exceptions and tailor solutions. Virtual teams will be able to collaborate, creating and experimenting in more spontaneous and responsive environments.

The transformation in business models draws a parallel with those sparked by the emergence of electricity. It took decades to move from lighting streets to creating the electric grid. The mass assembly line soon became commonplace, requiring an entirely new set of skills, management approaches and factory design. The United States was the first country to seize that opportunity and create an economy-wide impact with electricity. That helped the nation develop and lead subsequent innovations that became entirely new sectors: domestic appliances, the semiconductor industry, software and the Internet itself.

3

Internet of Things:
Security, Privacy and Safety

The Internet of Things (IoT) presents numerous benefits to consumers, and has the potential to change the ways in which consumers interact with technology in fundamental ways. In the future, the Internet of Things is likely to meld the virtual and physical worlds together in ways that are currently difficult to comprehend. From a security and privacy perspective, the predicted pervasive introduction of sensors and devices into currently intimate spaces – such as the home, the car, and with wearables and ingestible, even the body – poses particular challenges. As physical objects in our everyday lives increasingly detect and share observations about us, consumers will likely continue to want privacy.

IoT devices are poised to become more pervasive in our lives than mobile phones and will have access to the most sensitive personal data such as social security numbers and banking information, as their numbers are exponentially multiplied. A couple of security concerns on a single device such as a mobile phone can quickly turn to 50 or 60 concerns when considering multiple IoT devices in an interconnected home or business. In light of the importance of what IoT devices have access to, it is important to understand their security risk.

The Federal Trade Commission [6], has warned recently that the small size and limited processing power of many connected devices could limit the use of encryption and other security measures; it may also be difficult to patch flaws in low-cost and essentially disposable IoT devices.

The growth in these connected devices will spike over the next several years, according to numbers accumulated by Cisco Systems. What Cisco officials call the Internet of everything [7] will generate $19 trillion [8] in new revenues for businesses worldwide by 2020, and IDC analysts expect the IoT technology and services market to hit $8.9 trillion by the end of the decade [9].

However, while it may prove a financial boon for businesses and meet consumers' insatiable desire for more devices, the IoT will also increase the potential attack surface for hackers and other cyber-criminals. More devices online means more devices that need protecting [10], and IoT systems are not usually designed for cyberse-curity. The sophistication of cybercriminals is increasing, and the data breaches that are becoming increasingly familiar will only continue.

Internet of Things security is no longer a foggy future issue, as increasing number of such devices enter the market, and our lives, from self-parking cars to home automation systems to wearable smart devices. Google CEO Eric Schmidt [11] told world leaders at the World Economic Forum in Davos, Switzerland, in January 2015, "There will be so many sensors, so many devices, that you won't even sense it, it will be all around you." "It will be part of your presence all the time."

Issues around mobile security are already a challenge in this era of always connected devices. Think how much greater those challenges will be of a business has, for example, 10 IoT connected devices, and it is not going to get any easier. As the IoT evolves, there will be billions of connected devices – and each one represents a potential doorway into your IT infrastructure and your company or personal data.

3.1 IoT's Threats

We can list the threats of IoT under three categories: privacy, security and safety. Experts say the security threats of the Internet of Things are broad and potentially even crippling to systems. Since the IoT will have critical infrastructure components, it presents a good target for national and industrial espionage, as well as denial of service and other attacks. Another major area of concern is privacy with the personal information that will potentially reside on networks, also a likely target for cybercriminals.

One thing to keep in mind when evaluating security needs is that the IoT is still very much a work in progress. Many things are connected to the Internet now [12], and we will see an increase in this and the advent of contextual data sharing and autonomous machine actions based on that information, the IoT is the allocation of a virtual presence to a physical object, as it develops, and these virtual presences will begin to interact and exchange contextual information, [and] the devices will make decisions based on this contextual device. This will lead to very physical threats, around national infrastructure, possessions [e.g., cars and homes], environment, power, water and food supply, etc.

As a variety of objects become part of an interconnected environment, we have to consider that these devices have lost physical security, as they are going to be located in inhospitable environments, instantly accessible by the individual who is most motivated to tamper with the controls, attackers could potentially intercept, read or change data; they could tamper with control systems and change functionality, all adding to the risk scenarios.

3.2 Threats Are Real

Among the recent examples, one involves researchers who hacked into two cars and wirelessly disabled the brakes, turned the lights off and switched the brakes full on – all beyond the control of the driver. In another case, a luxury yacht was lured off course by researchers hacking the GPS signal that it was using for navigation.

Home control hubs have been found to be vulnerable, allowing attackers to tamper with heating, lighting, power and door locks, and other cases involve industrial control systems being hacked via their wireless network and sensors.

We are already seeing hacked TV sets and video cameras [and] child monitors that have raised privacy concerns and even hacked power meters, which to date have been used to steal electric power, adds Paul Henry, a principal at security consulting firm vNet Security LLC in Boynton Beach, FL, USA, and a senior instructor at the SANS Institute [13], a cooperative research and education organization in Bethesda, MD, USA. "A recent article spoke of a 'hacked light bulb,'" Henry says. "I can imagine a worm that would compromise large numbers of these Internet-connected devices and amass them in to a botnet of some kind. Remember it is not just the value or power of the device that the bad guy wants; it is the bandwidth it can access and use in a DDoS [distributed denial-of-service] attack."

The biggest concern, Henry says, is that the users of IoT devices will not regard the security of the devices they are connecting as being of great concern. "The issue is that the bandwidth of a compromised device can be used to attack a third party," he says. "Imagine a botnet of 100,000,000 IoT devices all making legitimate Web site requests on your corporate Web site at the same time."

Experts say the IoT will likely create unique and, in some cases, complex security challenges for organizations. As machines become autonomous, they are able to interact with other machines and make decisions that have an impact on the physical world. We have seen problems with automatic trading software, which can get trapped in a loop causing market drops. The systems may have failsafe built in, but these are coded by humans who are fallible, especially when they are writing code that works at the speed [and] frequency that computer programs can operate.

If a power system were hacked and they turned off the lights in an area of the city. No big deal perhaps for many, but for the thousands of people in the subway stations hundreds of feet underground in pitch darkness, the difference is massive. IoT allows the virtual world to interact with the physical world and that brings big safety issues.

3.3 What Can We Do

While threats will always exist with the IoT as they do with other technology endeavors, it is possible to bolster the security of IoT environments using security tools such as data encryption, strong user authentication, resilient coding and standardized and tested APIs that react in a predictable manner.

Some security tools will need to be applied directly to the connected devices. "The IoT and its cousin BYOD have the same security issues as traditional computers", says Randy Marchany, CISO at Virginia Tech University and the director of Virginia Tech's IT Security Laboratory [14]. "However, IoT devices usually don't have the capability to defend themselves and might have to rely on separate devices such as firewalls [and] intrusion detection/prevention systems. Creating a separate network segment is one option." "In fact, the lack of security tools on the devices themselves or a lack of timely security updates on the devices is what could make securing the IoT somewhat more difficult from other types of security initiatives", Marchany says. "Physical security is probably more of an issue, since these devices are usually out in the open or in remote locations and anyone can get physical access to it", says Marchany. "Once someone has physical access to the device, the security concerns rise dramatically."

"It doesn't help that vendors providing IoT technologies most likely have not designed security into their devices", says Marchany. "In the long term, IT executives should start requiring the vendors to assert [that] their products aren't vulnerable to common attacks such as those listed in the OWASP [15] [Open Web Application Security Project] Top 10 Web Vulnerabilities", he says. IT and security executives should "require vendors to list the vulnerabilities they know exist on their devices as part of the purchase process".

Security needs to be built in as the foundation of IoT systems, with rigorous validity checks, authentication and data verification, and all the data need to be encrypted. At the application level, software development organizations need to be better at writing code that is stable, resilient and trustworthy, with better code development standards, training, threat analysis and testing. As systems interact with each other, it is essential to have an agreed interoperability standard, which is safe and valid. Without a solid bottom-top structure, we will create more threats with every device added to the IoT. What we need is a secure and safe IoT with privacy-protected, tough trade-off but not impossible.

4

Internet of Things: More than Smart "Things"

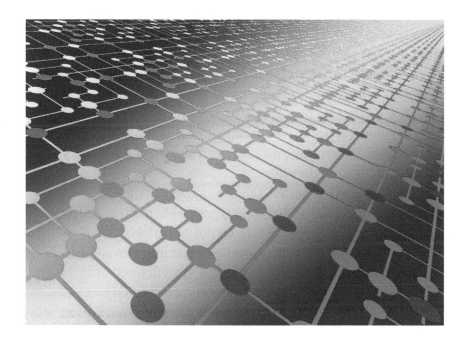

Experts forecast that by 2020, up to 28 billion devices will be connected to the Internet with only one-third of them being computers, smartphones and tablets. The remaining two-thirds will be other "devices" – sensors, terminals, household appliances, thermostats, televisions, automobiles, production machinery, urban infrastructure and many other "things", which traditionally have not been Internet-enabled.

This "Internet of Things" (IoT) represents a remarkable transformation of the way in which our world will soon interact. Much like the World Wide Web connected computers to networks, and the next evolution connected people to the Internet and other people, IoT looks poised to interconnect devices, people, environments, virtual objects and machines in ways that only science fiction writers could have imagined.

In a nutshell, the Internet of Things (IoT) is the convergence of connecting people, things, data and processes in transforming our life, business and everything in between.

A fair question to ask at this point is how IoT differs from machine to machine (M2M), which has been around for decades. Is IoT simply M2M with IPv6 addresses or is it really something revolutionary?

To answer this question, you need to know that M2M, built on proprietary and closed systems, was designed to move data securely in real time and mainly used for automation, instrumentation and control. It was targeted at point solutions (e.g., using sensors to monitor an oil well), deployed by technology buyers and seldom integrated with enterprise applications to help improve corporate performance.

On the contrary, IoT is built with interoperability in mind and is aimed at integrating sensor/device data with analytics and enterprise applications to provide unprecedented insights into business processes, operations and supplier and customer relationships. IoT is therefore a "tool" that is likely to become invaluable to CEOs, CFOs and General Managers of business units [16].

The technical definition of *The Internet of Things* (IoT) is the network of physical objects accessed through the Internet. These objects contain embedded technology to interact with internal states or the external environment.

In other words, when an object can sense and communicate, it changes how and where decisions are made and who makes them.

Due to the great breadth in the number of industries which have begun to be or soon will be affected by IoT, it is not right to define IoT as a unified "market". Rather, in an abstract sense, as a technology "wave" that will sweep across many industries at different points in time. The Internet of Things (IoT) is emerging as the third wave in the development of the Internet. The Internet wave of the 1990s connected 1 billion users, while the mobile wave of the 2000s connected another 2 billion. The IoT has the potential to connect 10 times as many (20 billion) "things" to the Internet by 2020, ranging from bracelets to cars [17].

Breakthroughs in the cost of sensors, processing power and band-width to connect devices are enabling ubiquitous connections cur-rently. Smart products like smart watches and thermostats (Nest) are already gaining traction as stated in the Goldman Sachs Global Investment Research report [1].

IoT has key attributes that distinguish it from the "regular" Inter-net, as captured by Goldman Sachs's S-E-N-S-E framework: *Sensing, Efficient, Networked, Specialized, Everywhere*. These attributes may change the direction of technology development and adoption, with significant implications for Tech companies (Figure 4.1).

4.1 Industrial Internet

"The Internet of Things will give IT managers a lot to think about", said Vernon Turner, Senior Vice President of Research at IDC [18]. "Enterprises will have to address every IT discipline to effectively balance the deluge of data from devices that are connected to the corporate network. In addition, IoT will drive tough organizational structure changes in companies to allow innovation to be transparent to everyone, while creating new competitive business models and products."

IoT is shaping modern business manufacturing to marketing. A lot has been already changed since the inception of the Internet, and many more will get changed with the greater Internet connectivity

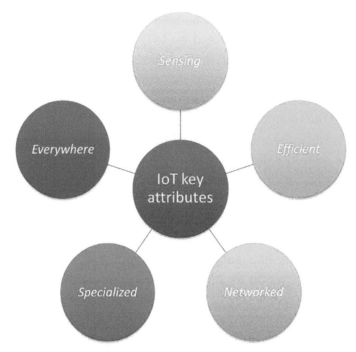

Figure 4.1 IoT Key Attributes.

and reach. The global network connecting people, data and machines transforming the modern business is also called Industrial Internet [19]. The so-called Industrial Internet has a potential of contributing $10 to $15 trillion to global GDP in the next two decades.

The "buzz" surrounding IoT has so far been more focused on the home, consumer and wearables markets, and tends to overshadow the enormous potential of Internet Protocol (IP) connected products in industrial and business/enterprise worlds. IoT in the consumer world is effectively a greenfield opportunity with no installed base and no dominant vendors, whereas there are many examples of connected products in this arena. The definition of the "Industrial and business/enterprise Internet" for IoT purposes refers to all non-consumer applications of the Internet of Things, ranging from smart cities, smart power grids, connected health, retail, supply chain and military applications. The technologies and solutions needed for creating smart connected products and processes share many common attributes across industrial and business verticals.

4.2 The IoT Value Chain

The IoT value chain is broad, extremely complex and spans many industries including those as diverse as semiconductors, industrial automation, networking, wireless and wireline operators, software vendors, security and systems integrators. Because of this complexity, very few companies will be able to successfully solve all of the associated problems or exploit the potential opportunities [16].

4.3 Internet of Things Predictions

According to IDC [16], IoT will go through a huge growth in the coming years in many directions:

1. *IoT and the cloud.* Within the next 5 years, more than 90% of all IoT data will be hosted on service provider platforms as cloud computing reduces the complexity of supporting IoT "Data Blending".
2. *IoT and security.* Within the next 2 years, 90% of all IT networks will have an IoT-based security breach, although many will be considered "inconveniences". Chief Information Security Officers (CISOs) will be forced to adopt new IoT policies.
3. *IoT at the edge.* By 2018, 40% of IoT-created data will be stored, processed, analyzed and acted upon close to, or at the edge of, the network.
4. *IoT and network capacity.* Within the next 3 years, 50% of IT networks will transition from having excess capacity to handle the additional IoT devices to being network constrained with nearly 10% of sites being overwhelmed.
5. *IoT and non-traditional infrastructure.* By 2017, 90% of datacenter and enterprise systems management will rapidly adopt

new business models to manage non-traditional infrastructure and BYOD device categories.

6. *IoT and vertical diversification.* Today, over 50% of IoT activity is centered in manufacturing, transportation, smart city and consumer applications, but within the next 5 years, all industries will have rolled out IoT initiatives.

7. *IoT and the smart city.* Competing to build innovative and sustainable smart cities, local government will represent more than 25% of all government external spending to deploy, manage and realize the business value of the IoT by 2018.

8. *IoT and embedded systems.* By 2018, 60% of IT solutions originally developed as proprietary, closed-industry solutions will become open-sourced, allowing a rush of vertical-driven IoT markets to form.

9. *IoT and wearables.* Within the next 5 years, 40% of wearables will have evolved into a viable consumer mass market alternative to smartphones.

10. *IoT and millennials.* By 2018, 16% of the population will be millennials and will be accelerating IoT adoption due to their reality of living in a connected world.

4.4 Challenges Facing IoT

IoT is shaping human life with greater connectivity and ultimate functionality through ubiquitous networking to the Internet. It will be more personal and predictive and merge the physical world and the virtual world to create a highly personalized and often predictive connected experience. With all the promises and potential, IoT still has to resolve three major issues [20]: unified standards for devices, privacy and security. Without the consideration of strong security at all joints of the IoT and protection of data, the progress of IoT will be hindered by litigation and social resistance. The expansion of IoT will be slow without common standards for the connected devices or sensors.

5

Internet of Things: Myths and Facts

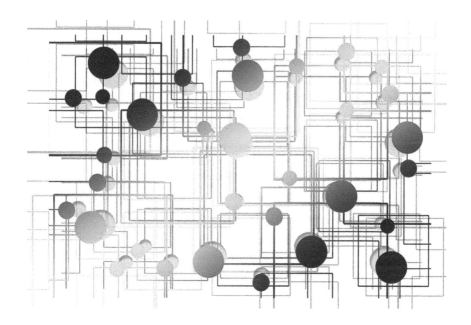

Any new technology involves a certain amount of uncertainty and business risk. In the case of the Internet of Things, however, many of the risks have been exaggerated or misrepresented. While the IoT vision will take years to mature fully, the building blocks to begin this process are already in place. Key hardware and software are either available today or under development; stakeholders need to address security and privacy concerns, collaborate to implement the open standards that will make the IoT safe, secure, reliable and interoperable, and allow the delivery of secured services as seamlessly as possible [21].

The Internet of Things (IoT) is a concept that describes a totally interconnected world. It is a world where devices of every shape and size are manufactured with "smart" capabilities that allow them to communicate and interact with other devices, exchange data, make autonomous decisions and perform useful tasks based on preset conditions. It is a world where technology will make life richer, easier, safer and more comfortable.

Cisco is expecting the industry to gross over $19 trillion in the next few years [22]. However, the problem is that these "things" have myths surrounding them, some of which are impacting how organizations develop the apps to support them.

5.1 IoT and Sensors

According to Cisco, "The fundamental problem posed by the Internet of Things is that network power remains very centralized [23]. Even in the era of the cloud, when you access data and services online you're mostly communicating with a relative few massive datacenters that might not be located particularly close to you. That works when you're not accessing a ton of data and when latency isn't a problem, but it doesn't work in the Internet of Things, where you could be doing something like monitoring traffic at every intersection in a city to more intelligently route cars and avoid gridlock. In that instance, if you had to wait for that chunk of data to be sent to a datacenter hundreds of miles away, processed, and then commands sent back to the streetlights, it would already be too late — the light would have already needed to change."

Cisco says that the solution is to do more computing closer to the sensors (fog computing [24]) that are gathering the data in the first place, so that the amount of data that needs to be sent to the centralized servers is minimized and the latency is mitigated. Cisco says that this data-crunching capability should be put on the router. This, however, is only part of the story. Getting the right data from the right device at the right time is not just about hardware and sensors, instead it is about data intelligence. If you can understand data and only distribute what is important, at the application level, this is more powerful than any amount of hardware you throw at the problem.

This prioritization of data should be done at the application level where there is logic. Combine this with data caching at the network edge and you have a solution that reduces latency.

5.2 IoT and Mobile Data

Smartphones certainly play a role in collecting some of this data and providing a user interface for accessing IoT applications, but they are ill-suited to play a more central role. Consider the example of home automation: It hardly makes sense for critical home-monitoring and security applications, such as those that protect an elderly resident against an accident or illness, to rely upon a smartphone as its decision-making hub. What happens when that person travels and his smartphone goes into airplane mode? Does his home security get interrupted, or home electricity shut down?

Such examples make it clear that the IoT will, with a few exceptions (such as "wearable" technology and bio-monitoring systems) and some automobile-related applications, rely mostly upon dedicated gateways and remote processing solutions – not on smartphones and mobile apps.

Today, without any IoT services, more than 80% of the traffic over LTE networks goes through Wi-Fi access points. What happens when that data increases by 22 times? In addition, cellular networks and communication devices have serious drawbacks in areas such as cost, power consumption, coverage and reliability.

So, will the Internet of Things have a place for smartphones and cellular communications? Absolutely, but in terms of performance, availability, cost, bandwidth, power consumption and other key attributes, the Internet of Things will require a much more diverse and innovative variety of hardware, software and networking solutions.

5.3 IoT and the Volume of Data

The IoT is going to produce a lot of data – an avalanche. As a result, some IoT experts believe that we will never be able to keep up with the ever-changing and ever-growing data being generated by the IoT because it is just not possible to monitor it all. Among all the data that is produced by the IoT, not all of it needs to be communicated to end-user apps such as real-time operational intelligence apps. This is because a lot of the chatter generated by devices is useless and does not represent any change in state. The apps are only interested in state changes, e.g. a light being on or off, a valve being open or shut and a traffic lane being open or closed. Rather than bombarding the apps with all of the device updates, apps should only be updated when the state changes.

5.4 IoT and Datacenters

Some argue that the datacenter is where all the magic happens for IoT. The datacenter is absolutely an important factor for the IoT; after all, this is where the data will be stored. But the myth here is that the datacenter is where the magic happens. What about the network? After all, IoT is nothing without the Internet actually supporting the distribution of information. So you might be able to store it or analyze it in a datacenter, but if the data cannot get there in the first place, is too slow in getting there or you cannot respond back in real time, then there is no IoT.

5.5 IoT is a Future Technology

The Internet of Things is simply the logical next step in an evolutionary process. The truth is that the technological building blocks of the

IoT – including microcontrollers, microprocessors, environmental and other types of sensors, and short-range and long-range networking communications – are in widespread use today. They have become far more powerful, even as they get smaller and less expensive to produce.

The Internet of Things, as we define it, while evolving the existing technologies further, simply adds one additional capability – a secured service infrastructure – to this technology mix. Such a service infrastructure will support the communication and remote control capabilities that enable a wide variety of Internet-enabled devices to work together [25].

5.6 IoT and Current Interoperability Standards

Everybody involved in the standards-making process knows that one size will not fit all – multiple (and sometimes overlapping) standards are a fact of life when dealing with evolving technology. At the same time, a natural pruning process will encourage stakeholders to standardize and focus on a smaller number of key standards. Standards issues pose a challenge, but these will be resolved as the standards process continues to evolve.

The Internet of Things will eventually include billions of interconnected devices. It will involve manufacturers from around the world and countless product categories. All of these devices must communicate, exchange data and perform closely coordinated tasks – and they must do so without sacrificing security or performance.

This sounds like a recipe for mass confusion. Fortunately, the building blocks to accomplish many of these tasks are already in place. Global standards bodies such as IEEE, International Society of Automation (ISA), the World Wide Web Consortium (W3C), and OMA SpecWorks (to name a few) bring together manufacturers, technology vendors, policy-makers and other interested stakeholders. As a result, while standards issues pose a short-term challenge for building the Internet of Things, the long-term process for resolving these challenges is already in place.

5.7 IoT and Privacy & Security

Security and privacy are major concerns – and addressing these concerns is a top priority. These are legitimate concerns. New technology often carries the potential for misuse and mischief, and it is vital to address the problem before it hinders personal privacy & security, innovation or economic growth. Manufacturers, standards organizations and policy-makers are already responding on several levels.

At the device level, security researchers are working on methods to protect embedded processors that, if compromised, would halt an attacker's ability to intercept data or compromise networked systems. At the network level, new security protocols will be necessary to ensure end-to-end encryption and authentication of sensitive data, and since with the Internet of Things the stakes are higher than the Internet, the industry is looking at full system-level security and optimization.

5.8 IoT and Limited Vendors

Open platforms and standards will create a base for innovation from companies of all types and sizes:

- *Open hardware architectures*. Open platforms are a proven way for developers and vendors to build innovative hardware with limited budgets and resources.
- *Open operating systems and software*. The heterogeneous nature of the Internet of Things will require a wide variety of software and applications, from embedded operating systems to Big Data analytics [26] and cross-platform development frameworks. Open software is extremely valuable in this context, since it gives developers and vendors the ability to adopt, extend and customize applications as they see fit – without onerous licensing fees or the risk of vendor lock-in.
- *Open standards*. As we discussed earlier, open standards and interoperability are vital to building the Internet of Things. An

environment where such a wide variety of devices and applications must work together simply cannot function unless it remains free from closed, proprietary standards.

Virtually all of the vendors, developers and manufacturers involved in creating the Internet of Things understand that open platforms will spur innovation and create rich opportunities for competition. Those that do not understand this may suffer the same fate as those that promoted proprietary networking standards during the Internet era: They were sidelined and marginalized.

5.9 Conclusion

The reality of the IoT is that if you want to distribute data from the "thing" across the network in real time over unreliable networks, you need intelligent data distribution. To lighten the load on the network by reducing your bandwidth usage, you need to understand your data. By understanding it, you can apply intelligence to only distribute what is relevant or what has changed. This means you send only small pieces of data across a congested network. The result is IoT apps with accurate, up-to date information, at scale, because you will be able to cope with the millions of devices connecting to your back end. You will not be hit with huge pieces of data at once, shutting down your services.

PART II

IoT Implementation & Standardization Challenges

6

Three Major Challenges Facing IoT

The Internet of Things (IoT) – a universe of connected things providing key physical data and further processing of that data in the cloud to deliver business insights – presents a huge opportunity for many players in all businesses and industries. Many companies are organizing themselves to focus on IoT and the connectivity of their future products and services. For the IoT industry to thrive, there are three categories of challenges to overcome, and this is true for any new trend in technology and not only IoT (Figure 6.1): technology, business and society [27, 28, 29].

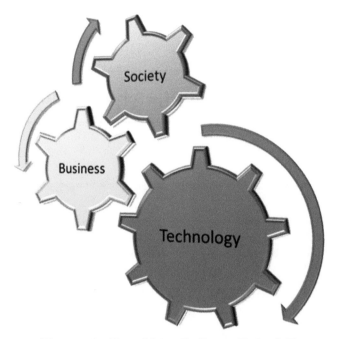

Figure 6.1 Three Major Challenges Facing IoT.

6.1 Technology

This part covers all technologies needed to make IoT systems function smoothly as a standalone solution or part of existing systems, and that is not an easy mission, there are many technological challenges (Figure 6.2), including security, connectivity, compatibility & longevity, standards and intelligent analysis & actions [30].

Figure 6.2 Technological Challenges Facing IoT.

6.2 Technological Challenges

Security: IoT has already turned into a serious security concern that has drawn the attention of prominent tech firms and government agencies across the world. The hacking of baby monitors, smart fridges, thermostats, drug infusion pumps, cameras and even the radio in your car are signifying a security nightmare being caused by the future of IoT. So many new nodes being added to networks and the Internet will provide malicious actors with innumerable attack vectors and possibilities to carry out their evil deeds, especially since a considerable number of them suffer from security holes.

The more important shift in security will come from the fact that IoT will become more ingrained in our lives. Concerns will no longer be limited to the protection of sensitive information and assets. Our very lives and health can become the target of IoT hack attacks [27].

There are many reasons behind the state of insecurity in IoT. Some of it has to do with the industry being in its "gold rush" state, where every vendor is hastily seeking to dish out the next innovative connected gadget before competitors do. Under such circumstances, functionality becomes the main focus and security takes a back seat.

Connectivity: Connecting so many devices will be one of the biggest challenges of the future of IoT, and it will defy the very structure of current communication models and the underlying technologies [28]. At present we rely on the centralized, server/client paradigm to authenticate, authorize and connect different nodes in a network.

This model is sufficient for current IoT ecosystems, where tens, hundreds or even thousands of devices are involved. But when networks grow to join billions and hundreds of billions of devices, centralized systems will turn into a bottleneck. Such systems will require huge investments and spending in maintaining cloud servers that can handle such large amounts of information exchange, and entire systems can go down if the server becomes unavailable.

The future of IoT will very much have to depend on decentralizing IoT networks. Part of it can become possible by moving some of the tasks to the edge, such as using fog computing models where smart devices such as IoT hubs take charge of mission-critical operations and cloud servers take on data gathering and analytical responsibilities [31].

Other solutions involve the use of peer-to-peer communications, where devices identify and authenticate each other directly and exchange information without the involvement of a broker. Networks will be created in meshes with no single point of failure. This model will have its own set of challenges, especially from a security perspective, but these challenges can be met with some of the emerging IoT technologies such as Blockchain [32].

Compatibility and Longevity: IoT is growing in many different directions, with many different technologies competing to become the standard. This will cause difficulties and require the deployment of extra hardware and software when connecting devices.

Other compatibility issues stem from non-unified cloud services, lack of standardized M2M protocols and diversities in firmware and operating systems among IoT devices.

Some of these technologies will eventually become obsolete in the next few years, effectively rendering the devices implementing them useless. This is especially important, since in contrast to generic computing devices, which have a life span of a few years, IoT appliances (such as smart fridges or TVs) tend to remain in service for much longer, and should be able to function even if their manufacturer goes out of service.

Standards: *Technology standards*, which include network protocols, communication protocols and data-aggregation standards are the sum of all activities of handling, processing and storing the data collected from the sensors [29]. This aggregation increases the value of data by increasing *the scale, scope, and frequency* of data available for analysis.

Challenges facing the adoptions of standards within IoT

- Standard for handling unstructured data: Structured data are stored in relational databases and queried through SQL, for example. Unstructured data are stored in different types of NoSQL databases without a standard querying approach.
- Technical skills to leverage newer aggregation tools: Companies that are keen on leveraging Big Data tools often face a shortage of talent to plan, execute and maintain systems.

Intelligent Analysis & Actions: The last stage in IoT implementation is extracting insights from data for analysis, where analysis is driven by *cognitive technologies* and the accompanying models that facilitate the use of cognitive technologies.

Factors driving adoption intelligent analytics within the IoT

- Artificial intelligence models can be improved with large data sets that are more readily available than ever before, thanks to the lower storage.
- Growth in crowdsourcing and open-source analytics software: Cloud-based crowdsourcing services are leading to new algorithms and improvements in existing ones at an unprecedented rate.
- Real-time data processing and analysis: Analytics tools such as complex event processing (CEP) enable processing and analysis of data on a real-time or a near-real-time basis, driving timely decision-making and action

Challenges facing the adoptions of intelligent analytics within IoT

- Inaccurate analysis due to flaws in the data and/or model: A lack of data or presence of outliers may lead to false positives or false negatives, thus exposing various algorithmic limitations.
- Legacy systems' ability to analyze unstructured data: Legacy systems are well suited to handle structured data; unfortunately, most IoT/business interactions generate unstructured data.

- Legacy systems' ability to manage real-time data: Traditional analytics software generally works on batch-oriented processing, wherein all the data are loaded in a batch and then analyzed.

The second phase of this stage is intelligent actions, which can be expressed as M2M and M2H interfaces, for example, with all the advancement in UI and UX technologies.

Factors driving adoption of intelligent actions within the IoT

- Lower machine prices
- Improved machine functionality
- Machines "influencing" human actions through behavioral-science rationale
- Deep learning tools

Challenges facing the adoption of intelligent actions within IoT

- Machines' actions in unpredictable situations
- Information security and privacy
- Machine interoperability
- Mean-reverting human behaviors
- Slow adoption of new technologies

6.3 Business

The bottom line is a big motivation for starting, investing in and operating any business. Without a sound and solid business model for IoT, we will have another bubble. This model must satisfy all the requirements for all kinds of e-commerce: vertical markets, horizontal markets and consumer markets. But this category is always a victim of regulatory and legal scrutiny.

End-to-end solution providers operating in vertical industries and delivering services using cloud analytics will be the most successful at monetizing a large portion of the value in IoT. While many IoT applications may attract modest revenue, some can attract more. For little burden on the existing communication infrastructure, operators have the potential to open up a significant source of new revenue using IoT technologies.

IoT can be divided into the following three categories, based on usage and clients base:

- **Consumer IoT** includes connected devices such as smart cars, phones, watches, laptops, connected appliances and entertainment systems.
- **Commercial IoT** includes things like inventory controls, device trackers and connected medical devices.
- **Industrial IoT** covers such things as connected electric meters, wastewater systems, flow gauges, pipeline monitors, manufacturing robots and other types of connected industrial devices and systems.

Categories of IoT

Clearly, it is important to understand the value chain and business model for the IoT applications for each category of IoT (Figure 6.3).

6.4 Society

Understanding IoT from the customer's and regulator's prospective is not an easy task for the following reasons:

- Customer demands and requirements change constantly.
- New uses for devices – as well as new devices – sprout and grow at breakneck speeds.
- Inventing and reintegrating must-have features and capabilities are expensive and take time and resources.
- The uses for Internet of Things technology are expanding and changing – often in uncharted waters.
- Consumer confidence: Each of these problems could put a dent in consumers' desire to purchase connected products, which would prevent the IoT from fulfilling its true potential.

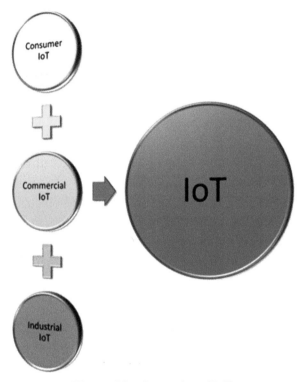

Figure 6.3   Categories of IoT.

- Lack of understanding or education by consumers of best practices for IoT devices security to help in improving privacy, for example, change default passwords of IoT devices.

6.5 Privacy

The IoT creates unique challenges to privacy, many that go beyond the data privacy issues that currently exist. Much of this stems from integrating devices into our environments without us consciously using them.

This is becoming more prevalent in consumer devices, such as tracking devices for phones and cars as well as smart televisions. In terms of the latter, voice recognition or vision features are being integrated that can continuously listen to conversations or watch for activity and selectively transmit that data to a cloud service for processing, which sometimes includes a third party. The collection of this information exposes legal and regulatory challenges facing data protection and privacy law.

In addition, many IoT scenarios involve device deployments and data collection activities with multinational or global scope that cross social and cultural boundaries. What will that mean for the development of a broadly applicable privacy protection model for the IoT?

In order to realize the opportunities of the IoT, strategies will need to be developed to respect individual privacy choices across a broad spectrum of expectations, while still fostering innovation in new technologies and services.

6.6 Regulatory Standards

Regulatory standards for data markets are missing especially for data brokers; they are companies that sell data collected from various sources. Even though data appear to be the currency of the IoT, there is a lack of transparency about who gets access to data and how those data are used to develop products or services and sold to advertisers and third parties. There is a need for clear guidelines on the retention, use and security of the data including metadata (the data that describe other data).

7

IoT Implementation and Challenges

The **Internet of Things (IoT)** is the network of physical objects – devices, vehicles, buildings and other items, which are embedded with electronics, software, sensors and network connectivity, which enables these objects to collect and exchange data. Implementing this concept is not an easy task by any measure for many reasons, including the complex nature of the different components of the ecosystem of IoT.

Figure 7.1 Components of IoT implementation.

To understand the gravity of this task, we will explain all the five components of IoT Implementation (Figure 7.1).

7.1 Components of IoT Implementation

- Sensors
- Networks
- Standards
- Intelligent Analysis
- Intelligent Actions

7.1.1 Sensors

According to IEEE, sensors can be defined as an electronic device that produces electrical, optical or digital data derived from a physical condition or event. Data produced from sensors is then electronically transformed, by another device, into information (output) that is useful in decision-making done by "intelligent" devices or individuals (people) [33].

Types of Sensors: Active sensors and passive sensors.

The selection of sensors is greatly impacted by many factors, including:

- Purpose (temperature, motion, bio, etc.)
- Accuracy
- Reliability
- Range
- Resolution
- Level of intelligence (dealing with noise and interference)

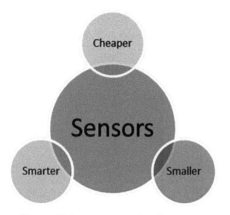

Figure 7.2   New trends of sensors.

The driving forces for using sensors in IoT today are new trends in technology that made sensors cheaper, smarter and smaller (Figure 7.2).

Challenges facing IoT sensors:

- Power consumption
- Security
- Interoperability

7.1.2 Networks

The second step of this implantation is to transmit the signals collected by sensors over networks with all the different components of a typical network including routers, bridges in different topologies, including LAN, MAN and WAN. Connecting the different parts of networks to the sensors can be done by different technologies, including Wi-Fi, Bluetooth, Low Power Wi-Fi, Wi-Max, regular Ethernet, Long Term Evolution (LTE) and the recent promising technology of Li-Fi (using light as a medium of communication between the different parts of a typical network including sensors).

The driving forces for widespread network adoption in IoT can be summarized as follows:

- High data rate
- Low prices of data usage
- Virtualization (X-Defined Network trends)
- XaaS concept (SaaS, PaaS and IaaS)
- IPv6 deployment

Challenges facing network implementation in IoT

- The enormous growth in the number of connected devices
- Availability of networks coverage
- Security
- Power consumption

7.1.3 Standards

The third stage in the implementation process involves all activities of handling, processing and storing the data collected from the sensors. This aggregation increases the value of data by increasing *the scale, scope and frequency* of data available for analysis but aggregation is only achieved through the use of various standards depending on the IoT application used.

Types of Standards

Two types of standards relevant for the aggregation process: *technology standards* (including network protocols, communication protocols and data aggregation standards) and *regulatory standards* (related to security and privacy of data, among other issues).

Technology Standards

- Network protocols (e.g., Wi-Fi)
- Communications protocols (e.g., HTTP)
- Data aggregation standards (e.g., extraction, transformation, loading (ETL)).

Regulatory Standards

Set and administrated by government agencies like FTC, for example, Fair Information Practice Principles (FIPP) and US Health Insurance Portability and Accountability Act (HIPAA), just to mention a few.

Challenges facing the adoptions of standards within IoT

- **Standard for handling unstructured data:** Structured data are stored in relational databases and queried through SQL. Unstructured data are stored in different types of NoSQL databases without a standard querying approach.
- **Security and privacy issues:** There is a need for clear guidelines on the retention, use and security of the data as well as metadata (the data that describes other data).
- **Regulatory standards for data markets:** Data brokers are companies that sell data collected from various sources. Even though data appear to be the *currency* of the IoT, there is a lack of transparency about who gets access to data and how those data are used to develop products or services and sold to advertisers and third parties.
- **Technical skills to leverage newer aggregation tools:** Companies that are keen on leveraging Big Data tools often face a shortage of talent to plan, execute and maintain systems.

7.1.4 Intelligent Analysis

The fourth stage in IoT implementation is extracting insight from data for analysis, Analysis is driven by *cognitive technologies* and the accompanying models that facilitate the use of cognitive technologies.

With advances in cognitive technologies' ability to process varied forms of information, vision and voice have also become usable. Below is a list of selected cognitive technologies that are experiencing increasing adoption and can be deployed for predictive and prescriptive analytics:

- **Computer vision** refers to computers' ability to identify objects, scenes and activities in images.
- **Natural-language processing** refers to computers' ability to work with text the way humans do, extracting meaning from text or even generating text.
- **Speech recognition** focuses on accurately transcribing human speech.

Factors driving adoption intelligent analytics within the IoT

- **Artificial intelligence models** can be improved with large data sets that are more readily available than ever before, thanks to the lower storage.
- **Growth in crowdsourcing and open-source analytics software:** Cloud-based crowdsourcing services are leading to new algorithms and improvements in existing ones at an unprecedented rate.
- **Real-time data processing and analysis:** Analytics tools such as complex event processing (CEP) enable processing and analysis of data on a real-time or a near-real-time basis, driving timely decision-making and action.

Challenges facing the adoptions of intelligent analytics within IoT

- **Inaccurate analysis due to flaws in the data and/or model:** A lack of data or presence of outliers may lead to false positives or false negatives, thus exposing various algorithmic limitations.
- **Legacy systems' ability to analyze *unstructured data*:** Legacy systems are well suited to handle ***structured data***; unfortunately, most IoT/business interactions generate unstructured data.
- **Legacy systems' ability to manage real-time data:** Traditional analytics software generally works on batch-oriented processing, wherein all the data are loaded in a batch and then analyzed.

7.1.5 Intelligent Actions

Intelligent actions can be expressed as M2M and M2H interface, for example, with all the advancement in UI and UX technologies.

Factors driving adoption of intelligent actions within the IoT

- Lower machine prices
- Improved machine functionality
- Machines "influencing" human actions through behavioral-science rationale
- Deep learning tools

Challenges facing the adoption of intelligent actions within IoT

- Machines' actions in unpredictable situations
- Information security and privacy
- Machine interoperability
- Mean-reverting human behaviors
- Slow adoption of new technologies

The Internet of Things (IoT) is an ecosystem of ever-increasing complexity, it is the next wave of innovation that will humanize every object in our life, which is the next level to automating every object in our life. Convergence of technologies will make IoT implementation much easier and faster, which in turn will improve many aspects of our life at home and at work and in between [34].

8

IoT Standardization and Implementation Challenges

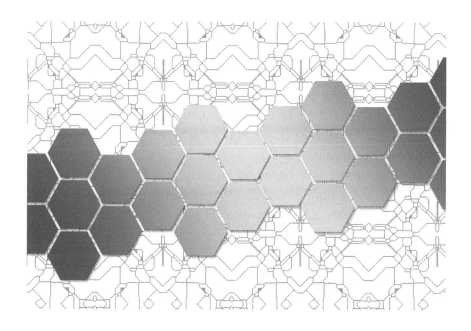

The rapid evolution of the IoT market has caused an explosion in the number and variety of IoT solutions. Additionally, large amounts of funding are being deployed at IoT startups. Consequently, the focus of the industry has been on manufacturing and producing the right types of hardware to enable those solutions. In the current model, most IoT solution providers have been building all components of the stack,

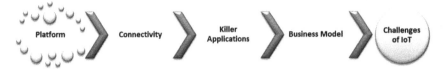

Figure 8.1   Hurdles facing IoT standardization.

from the hardware devices to the relevant cloud services or as they would like to name it as "IoT solutions", as a result, there is a lack of consistency and standards across the cloud services used by the different IoT solutions.

As the industry evolves, the need for a standard model to perform common IoT backend tasks, such as processing, storage and firmware updates, is becoming more relevant. In that new model, we are likely to see different IoT solutions work with common backend services, which will guarantee levels of interoperability, portability and manageability that are almost impossible to achieve with the current generation of IoT solutions.

Creating that model will never be an easy task by any level of imagination. There are hurdles and challenges facing the standardization and implementation of IoT solutions (Figure 8.1), and that model needs to overcome all of them [38].

8.1 IoT Standardization

The hurdles facing IoT standardization can be divided into four categories: platform, connectivity, business model and killer applications.

- **Platform:** This part includes the form and design of the products (UI/UX), analytics tools used to deal with the massive data streaming from all products in a secure way and scalability, which means wide adoption of protocols like IPv6 in all vertical and horizontal markets is needed.
- **Connectivity:** This phase includes all parts of the consumer's routine, from using wearables, smart cars, smart homes and, in the

big scheme, smart cities. From the business perspective, we have connectivity using IIoT (Industrial Internet of Things), where M2M communications dominate the field.

- **Killer applications:** In this category, there are three functions needed to have killer applications: control "things", collect "data" and analyze "data". IoT needs killer applications to drive the business model using a unified platform.
- **Business model:** The bottom line is a big motivation for starting, investing in and operating any business. Without a sound and solid business model for IoT, we will have another bubble, and this model must satisfy all the requirements for all kinds of e-commerce: vertical markets, horizontal markets and consumer markets. But this category is always a victim of regulatory and legal scrutiny.

All four categories are inter-related, and you need to make all of them work. Missing one will break that model and stall the standardization process. A lot of work is needed in this process, and many companies are involved in each of one of the categories. Bringing them to the table to agree on a unifying model will be a daunting task.

8.2 IoT Implementation

The second part of the model is IoT implementations; implementing IoT is not an easy process by any measure for many reasons, including the complex nature of the different components of the ecosystem of IoT. To understand the significance of this process, we will explore all the **five** components of IoT implementation: sensors, networks, standards, intelligent analysis and intelligent actions (Figure 8.2).

8.2.1 Sensors

There two types of sensors: active sensors and passive sensors. *The driving forces for using sensors in IoT* today are new trends in technology that made sensors cheaper, smarter and smaller. But the

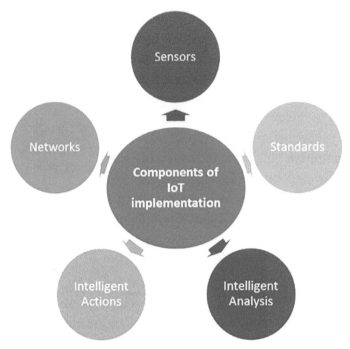

Figure 8.2   Components of IoT implementations.

challenges facing IoT sensors are power consumption, security and interoperability.

8.2.2 Networks

The second component of IoT implantation is to transmit the signals collected by sensors over networks with all the different components of a typical network including routers and bridges in different topologies. Connecting the different parts of networks to the sensors can be done by different technologies, including Wi-Fi, Bluetooth, Low Power Wi-Fi, Wi-Max, regular Ethernet, Long Term Evolution (LTE) and the recent promising technology of Li-Fi (using light as a medium of communication between the different parts of a typical network including sensors) [36].

The driving forces for widespread network adoption in IoT are high data rate, low prices of data usage, virtualization (X-Defined Network trends), XaaS concept (SaaS, PaaS and IaaS) and IPv6 deployment. But the *challenges facing network implementation in IoT are* the enormous growth in a number of connected devices, availability of networks coverage, security and power consumption.

8.2.3 Standards

The third stage in the implementation process includes the sum of all activities of handling, processing and storing the data collected from the sensors. This aggregation increases the value of data by increasing *the scale, scope and frequency* of data available for analysis, but aggregation is only achieved through the use of various standards depending on the IoT application used.

There are two types of standards relevant for the aggregation process: *technology standards* (including network protocols, communication protocols and data aggregation standards) and *regulatory standards* (related to security and privacy of data, among other issues). *Challenges facing the adoptions of standards within IoT are* standard for handling unstructured data, security and privacy issues in addition to regulatory standards for data markets [35].

8.2.4 Intelligent Analysis

The fourth stage in IoT implementation is extracting insight from data for analysis. IoT analysis is driven by *cognitive technologies* and the accompanying models that facilitate the use of cognitive technologies. With advances in cognitive technologies, the ability to process varied forms of information, vision and voice has also become usable and opened the doors for an in-depth understanding of the non-stop streams of real-time data. *Factors driving adoption intelligent analytics within the IoT* are: artificial intelligence models, growth in crowdsourcing and open-source analytics software, real-time data processing and analysis. *Challenges facing the adoption of analytics within IoT* are: inaccurate analysis due to flaws in the data and/or model, legacy systems' ability to analyze *unstructured data and* legacy systems' ability to manage real-time data [35].

8.2.5 Intelligent Actions

Intelligent actions can be expressed as M2M (Machine to Machine) and M2H (Machine to Human) interfaces, for example, with all the advancement in UI and UX technologies. *Factors driving adoption of intelligent actions within the IoT* are: lower machine prices, improved machine functionality, machines "influencing" human actions through behavioral-science rationale and Deep Learning tools. *Challenges facing the adoption of intelligent actions within IoT* are: machines' actions in unpredictable situations, information security and privacy, machine interoperability, mean-reverting human behaviors and slow adoption of new technologies.

8.3 The Road Ahead

The Internet of Things (IoT) is an ecosystem of ever-increasing complexity. It is the next wave of innovation that will humanize every object in our life, and is the next level to automating every object in our life and convergence of technologies that will make IoT implementation much easier and faster, which in turn will improve many aspects of our life at home and at work and in between. From refrigerators to parking spaces to houses, IoT is bringing more and more things into the digital fold every day, which will likely make IoT a multi-trillion-dollar industry in the near future. One possible outcome of successful standardization of IoT is the implementation of "IoT as a Service" technology if that service offered and used the same way we use other flavors of "as a service" technologies. Today, the possibilities of applications in real life are unlimited. But we have a long way to achieve that dream, and we need to overcome many obstacles and barriers in two fronts: consumers and businesses before we can harvest the fruits of such technology [37, 39].

9

Challenges Facing IoT Analytics Success

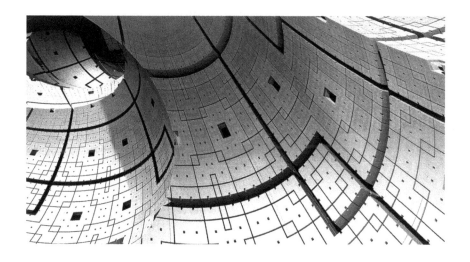

The Internet of Things (IoT) is an ecosystem of ever-increasing complexity; it is the next wave of innovation that will humanize every object in our life. IoT is bringing more and more devices (things) into the digital fold every day, which will likely make IoT a multi-trillion-dollars industry in the near future. To understand the scale of interest in the Internet of Things (IoT), just check how many conferences, articles and studies have been conducted about IoT recently, including a recent well-written article about IoT future by SAP listing 21 experts' insights into IoT future [40]. This interest has hit fever pitch point last year as many companies see big opportunity and believe that IoT holds the promise to expand and improve businesses processes and accelerate growth.

However, the rapid evolution of the IoT market has caused an explosion in the number and variety of IoT solutions, which created real challenges as the industry evolves, mainly, the urgent need for a reliable IoT model to perform common tasks such as sensing, processing, storage and communicating. Developing that model will never be an easy task by any stretch of the imagination; there are many hurdles and challenges facing a real reliable IoT model.

One of the crucial functions of using IoT solutions is to take advantage of IoT analytics to exploit the information collected by "things" in many ways, for example, to understand customer behavior, to deliver services, to improve products and to identify and intercept business moments. IoT demands new analytic approaches as data volumes increase through 2021 to astronomical levels, and the needs of the IoT analytics may diverge further from traditional analytics.

There are many challenges facing IoT analytics (Figure 9.1), including *Data Structures, Combing Multi Data Formats, The Need*

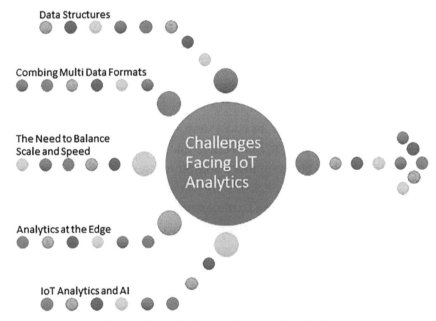

Figure 9.1   Challenges facing IoT analytics.

to Balance Scale and Speed, Analytics at the Edge and IoT Analytics and AI.

9.1 Data Structures

Most sensors send out data with a time stamp and most of the data is "boring" with nothing happening for much of the time. However, once in a while, something serious happens and needs to be attended to. While static alerts based on thresholds are a good starting point for analyzing this data, they cannot help us advance to diagnostic or predictive or prescriptive phases. There may be relationships between data pieces collected at specific intervals of time, in other words, classic time series challenges.

9.2 Combining Multiple Data Formats

While time series data have established techniques and processes for handling the insights that would really matter cannot come from sensor data alone. There are usually strong correlations between sensor data and other unstructured data. For example, a series of control unit fault codes may result in a specific service action that is recorded by a mechanic. Similarly, a set of temperature readings may be accompanied by a sudden change in the macroscopic shape of a part that can be captured by an image or change in the audible frequency of a spinning shaft. We would need to develop techniques where structured data must be effectively combined with unstructured data or what we call Dark Data [41].

9.3 The Need to Balance Scale and Speed

Most of the serious analysis for IoT will happen in the cloud, a data center, or more likely a hybrid cloud and server-based environment. This is because, despite the elasticity and scalability of the cloud, it may not be suited for scenarios requiring large amounts of data to be processed in real time. For example, moving 1 terabyte over a 10 Gbps network takes 13 minutes, which is fine for batch processing and management of historical data, but it is not practical for analyzing real-time event streams; a recent example is data transmitted by autonomous cars especially in critical situations that required a split second decision.

At the same time, because different aspects of IoT analytics may need to scale more than others, the analysis algorithm implemented should support flexibility whether the algorithm is deployed in the edge, data center, or cloud.

9.4 IoT Analytics at the Edge

IoT sensors, devices and gateways are distributed across different manufacturing floors, homes, retail stores and farm fields, to name just a few locations. Yet moving 1 terabyte of data over a 10 Mbps broadband network will take 9 days. Therefore, enterprises need to plan on how to address the projected 40% of IoT data that will be processed at the edge in just a few years' time [42]. This is particularly true for large IoT deployments, where billions of events may stream through each second, but systems only need to know an average over time or be alerted when a trends fall outside established parameters.

The answer is to conduct some analytics on IoT devices or gateways at the edge and send aggregated results to the central system. Through such edge analytics, organizations can ensure the timely detection of important trends or aberrations while significantly reducing network traffic to improve performance.

Performing edge analytics requires very lightweight software, since IoT nodes and gateways are low-power devices with limited strength for query processing. To deal with this challenge, Fog Computing is the champion [31].

Fog computing allows computing, decision-making and action-taking to happen via IoT devices and only pushes relevant data to the cloud. Cisco coined the term "Fog computing" and gave a brilliant definition for it: "The fog extends the cloud to be closer to the things that produce and act on IoT data. These devices, called fog nodes, can be deployed anywhere with a network connection: on a factory floor, on top of a power pole, alongside a railway track, in a vehicle, or on an oil rig. Any device with computing, storage, and network connectivity can be a fog node. Examples include industrial controllers, switches, routers, embedded servers, and video surveillance cameras." The major benefits of using fog computing are: minimizing latency, conserving network bandwidth and addressing security concerns at all levels of the network. In addition, it operates reliably with quick decisions, collects and secures wide range of data, moves data to the best place for processing, lowers expenses of using high computing power only when needed and less bandwidth and gives better analysis and insights of local data.

Keep in mind that fog computing is not a replacement of cloud computing by any measures, it works in conjunction with cloud computing, optimizing the use of available resources. But it was the product of a need to address many challenges: real-time process and action of incoming data, as well as limitation of resources like bandwidth and computing power, another factor helping fog computing is the fact that it takes advantage of the distributed nature of today's virtualized IT resources. This improvement to the data-path hierarchy is enabled by the increased compute functionality that manufacturers are building into their edge routers and switches.

9.5 IoT Analytics and AI

The greatest – and as yet largely untapped – power of IoT analysis is to go beyond reacting to issues and opportunities in real time and instead prepare for them beforehand. That is why, prediction is central to many IoT analytics strategies, whether to project demand, anticipate maintenance, detect fraud, predict churn or segment customers.

Artificial intelligence (AI) uses and improves current statistical models for handling prediction. AI will automatically learn underline rules, providing an attractive alternative to rules-only systems, which

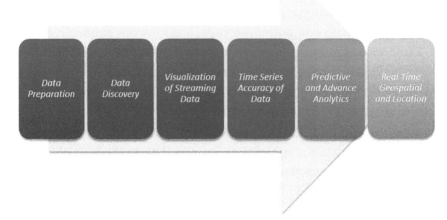

Figure 9.2   AI and IoT Data Analysis.

require professionals to author rules and evaluate their performance. When AI is applied, it provides valuable and actionable insights.

There are six types of IoT Data Analysis where AI can help (Figure 9.2):

1. *Data Preparation*: Defining pools of data and cleaning them, which will take us to concepts like Dark Data and Data Lakes [43].
2. *Data Discovery*: Finding useful data in the defined pools of data.
3. *Visualization of Streaming Data*: On the fly dealing with streaming data by defining, discovering and visualizing data in smart ways to make it easy for the decision-making process to take place without delay.
4. *Time Series Accuracy of Data*: Keeping the level of confidence in data collected high with high accuracy and integrity of data.
5. *Predictive and Advance Analytics*: Very important step where decisions can be made based on data collected, discovered and analyzed.
6. *Real-Time Geospatial and Location (logistical Data)*: Maintaining the flow of data smooth and under control.

But it is not all "nice & rosy, comfy and cozy", as there are challenges in using AI in IoT [44], such as compatibility, complexity, privacy/security/safety, ethical and legal issues and artificial stupidity.

Many IoT ecosystems will emerge, and commercial and technical battles between these ecosystems will dominate areas such as the smart home, the smart city, financials and healthcare. But the real winners will be the ecosystems with better, reliable, fast and smart IoT Analytics tools, after all what matters is how can we change data to insights and insights to actions and actions to profit.

PART III

Securing IoT

10

How to Secure the Internet of Things

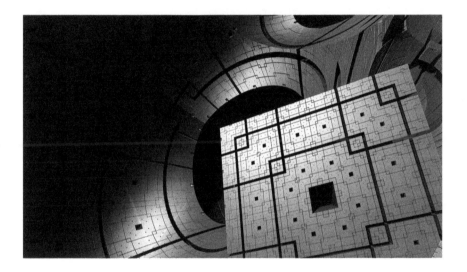

The Internet of Things (IoT) as a concept is fascinating and exciting, but the key to gaining real business value from it is effective communication between all elements of the architecture so you can deploy applications faster, process and analyze data at lightning speeds and make decisions as soon as you can.

IoT architecture can be represented by four systems (Figure 10.1):

1. **Things:** These are defined as uniquely identifiable nodes, primarily sensors that communicate without human interaction using IP connectivity.

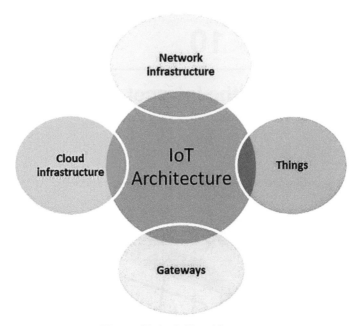

Figure 10.1   IoT architecture.

2. **Gateways:** These act as intermediaries between things and the cloud to provide the needed Internet connectivity, security and manageability.

3. **Network infrastructure:** This is composed of routers, aggregators, gateways, repeaters and other devices that control data flow.

4. **Cloud infrastructure:** Cloud infrastructure contains large pools of virtualized servers and storage that are networked together.

Next-generation trends [45], namely social networks, Big Data, cloud computing and mobility, have made many things possible that were not just a few years ago. Add to that, the convergence of global trends and events that are fueling today's technological advances and enabling innovation including:

- Efficiency and cost-reduction initiatives in key vertical market
- Government incentives encouraging investment in these new technologies

- Lower manufacturing costs for smart devices
- Reduced connectivity costs
- More efficient wired and wireless communications
- Expanded and affordable mobile networks

Internet of Things (IoT) is one big winner in this entire ecosystem. IoT is creating new opportunities and providing a competitive advantage for businesses in current and new markets. It touches everything – not just the data, but how, when, where and why you collect it. The technologies that have created the Internet of Things are not changing the Internet only, but rather change the things connected to the internet – the devices and gateways on the edge of the network that are now able to request a service or start an action without human intervention at many levels.

Because the generation and analysis of data is so essential to the IoT, consideration must be given to protecting data throughout its life cycle. Managing information at this level is complex because data will flow across many administrative boundaries with different policies and intents. Generally, data is processed or stored on edge devices that have highly limited capabilities and are vulnerable to sophisticated attacks.

Given the various technological and physical components that truly make up an IoT ecosystem, it is good to consider the IoT as a system of systems. The architecting of these systems that provide business value to organizations will often be a complex task, as enterprise architects work to design integrated solutions that include edge devices [46], applications, transports, protocols and analytics capabilities that make up a fully functioning IoT system. This complexity introduces challenges to keep the IoT secure and ensure that a particular instance of the IoT cannot be used as a jumping off point to attack other enterprise information technology (IT) systems.

International Data Corporation (IDC) reported that 90% of organizations that implement the IoT suffered an IoT-based breach of back-end IT systems in 2017 [47].

10.1 Challenges to Secure IoT Deployments

Regardless of the role your business has within the Internet of Things ecosystem – device manufacturer, solution provider, cloud provider, systems integrator or service provider – you need to know how to get the greatest benefit from this new technology that offers such highly diverse and rapidly changing opportunities.

Handling the enormous volume of existing and projected data is daunting. Managing the inevitable complexities of connecting to a seemingly unlimited list of devices is complicated. And the goal of turning the deluge of data into valuable actions seems impossible because of the many challenges. The existing security technologies will play a role in mitigating IoT risks, but they are not enough. The goal is to get data securely at the right place, at the right time and in the right format, and it is easier said than done for many reasons. Cloud Security Alliance (CSA) [48], in a recent report, listed some of the challenges:

- Many IoT systems are poorly designed and implemented, using diverse protocols and technologies that create complex configurations
- Lack of mature IoT technologies and business processes
- Limited guidance for life cycle maintenance and management of IoT devices
- The IoT introduces unique physical security concerns
- IoT privacy concerns are complex and not always readily evident
- Limited best practices available for IoT developers
- There is a lack of standards for authentication and authorization of IoT edge devices
- There are no best practices for IoT-based incident response activities
- Audit and logging standards are not defined for IoT components
- Restricted interfaces available IoT devices to interact with security devices and applications

- No focus yet on identifying methods for achieving situational awareness of the security posture of an organization's IoT assets
- Security standards, for platform configurations, involving virtualized IoT platforms supporting multi-tenancy is immature
- Customer demands and requirements change constantly
- New uses for devices – as well as new devices – sprout and grow at breakneck speeds
- Inventing and reintegrating must-have features and capabilities are expensive and consume time and resources
- The uses for Internet of Things technology are expanding and changing – often in uncharted waters
- Developing the embedded software that provides Internet of Things value can be difficult and expensive.

Some real examples of threats and attack vectors that malicious actors could take advantage of are:

- Control systems, vehicles and even the human body can be accessed and manipulated causing injury or worse
- Healthcare providers can improperly diagnose and treat patients
- Intruders can gain physical access to homes or commercial businesses
- Loss of vehicle control
- Safety critical information such as warnings of a broken gas line can go unnoticed
- Critical infrastructure damage
- Malicious parties can steal identities and money
- Unanticipated leakage of personal or sensitive information
- Unauthorized tracking of people's locations, behaviors and activities
- Manipulation of financial transactions
- Vandalism, theft or destruction of IoT assets
- Ability to gain unauthorized access to IoT devices
- Ability to impersonate IoT devices.

10.2 Dealing with the Challenges and Threats

Gartner [49] predicted at its security and risk management summit in Mumbai, India, this year, that more than 20% of businesses will have deployed security solutions for protecting their IoT devices and services by 2017. IoT devices and services will expand the surface area for cyber-attacks on businesses, by turning physical objects that used to be offline into online assets communicating with enterprise networks. Businesses will have to respond by broadening the scope of their security strategy to include these new online devices.

Businesses will have to tailor security to each IoT deployment according to the unique capabilities of the devices involved and the risks associated with the networks connected to those devices. BI Intelligence expects spending on solutions to secure IoT devices and systems to increase five fold over the next 4 years.

10.3 The Optimum Platform

Developing solutions for the Internet of Things requires unprecedented collaboration, coordination and connectivity for each piece in the system, and throughout the system as a whole. All devices must work together and be integrated with all other devices, and all devices must communicate and interact seamlessly with connected systems and infrastructures. It is possible, but it can be expensive, time consuming and difficult.

The optimum platform for IoT can:

- Acquire and manage data to create a standards-based, scalable and secure platform
- Integrate and secure data to reduce cost and complexity while protecting your investment
- Analyze data and act by extracting business value from data and then acting on it.

10.4 Last Word

Security needs to be built in as the foundation of IoT systems, with rigorous validity checks, authentication, data verification and all the data needs to be encrypted. At the application level, software development organizations need to be better at writing code that is stable, resilient and trustworthy, with better code development standards, training, threat analysis and testing. As systems interact with each other, it is essential to have an agreed interoperability standard, which is safe and valid. Without a solid bottom–top structure, we will create more threats with every device added to the IoT. What we need is a secure and safe IoT with privacy-protected, tough trade-off, but it is not impossible.

11

Using Blockchain to Secure IoT

In an IoT world, information is the "fuel" that is used to change the physical state of environments through devices that are not general-purpose computers but, instead, devices and services that are designed for specific purposes. As such, the IoT is at a conspicuous inflection point for IT security [49].

11.1 Challenges to Secure IoT Deployments

Regardless of the role, your business has within the Internet of Things ecosystem – device manufacturer, solution provider, cloud provider, systems integrator or service provider – you need to know how to get the greatest benefit from this new technology that offers such highly diverse and rapidly changing opportunities.

Handling the enormous volume of existing and projected data is daunting. Managing the inevitable complexities of connecting to a seemingly unlimited list of devices is complicated. And the goal of turning the deluge of data into valuable actions seems impossible due to the many challenges. The existing security technologies will play a role in mitigating IoT risks, but they are not enough. The goal is to get data securely at the right place, at the right time and in the right format, and it is easier said than done for many reasons.

11.2 Dealing with the Challenges and Threats

Gartner reported that more than 20% of businesses need to deploy security solutions for protecting their IoT devices and services. IoT devices and services will expand the surface area for cyber-attacks on businesses, by turning physical objects that used to be offline into online assets communicating with enterprise networks. Businesses will have to respond by broadening the scope of their security strategy to include these new online devices.

Businesses will have to tailor security to each IoT deployment according to the unique capabilities of the devices involved and the risks associated with the networks connected to those devices. BI Intelligence expects spending on solutions to secure IoT devices and systems to increase five fold over the next 4 years.

11.3 The Optimum Platform

Developing solutions for the Internet of Things requires unprecedented collaboration, coordination and connectivity for each piece in the system, and throughout the system as a whole. All devices must work together and be integrated with all other devices, and all devices must communicate and interact seamlessly with connected systems and infrastructures in a secure way. It is possible, but it can be expensive, time-consuming, and difficult unless the new line of thinking and a new approach to IoT security emerged away from the current centralized model.

The current IoT ecosystems rely on centralized, brokered communication models, otherwise known as the server/client paradigm. All devices are identified, authenticated and connected through cloud servers that sport huge processing and storage capacities. The connection between devices will have to exclusively go through the Internet, even if they happen to be a few feet apart.

While this model has connected generic computing devices for decades and will continue to support small-scale IoT networks as we see them today, it will not be able to respond to the growing needs of the huge IoT ecosystems of tomorrow.

Existing IoT solutions are expensive because of the high infrastructure and maintenance cost associated with centralized clouds, large server farms and networking equipment. The sheer amount of communications that will have to be handled when IoT devices grow to tens of billions will increase those costs substantially.

Even if the unprecedented economical and engineering challenges are overcome, cloud servers will remain a bottleneck and point of failure that can disrupt the entire network. This is especially important as more critical tasks.

Moreover, the diversity of ownership of devices and their supporting cloud infrastructure makes machine-to-machine (M2M) communications difficult. There is no single platform that connects all devices and no guarantee that cloud services offered by different manufacturers are interoperable and compatible.

11.4 Decentralizing IoT Networks

A decentralized approach to IoT networking would solve many of the questions above. Adopting a standardized peer-to-peer communication model to process the hundreds of billions of transactions between devices will significantly reduce the costs associated with installing and maintaining large centralized data centers and will distribute computation and storage needs across the billions of devices that form IoT networks. This will prevent failure in any single node in a network from bringing the entire network to a halting collapse [50].

However, establishing peer-to-peer communications will present its own set of challenges, chief among them the issue of security. And as we all know, IoT security is much more than just about protecting sensitive data. The proposed solution will have to maintain privacy and security in huge IoT networks and offer some form of validation and consensus for transactions to prevent spoofing and theft.

To perform the functions of traditional IoT solutions without a centralized control, any decentralized approach must support three fundamental functions:

- Peer-to-peer messaging
- Distributed file sharing
- Autonomous device coordination

11.5 The Blockchain Approach

Blockchain, the "distributed ledger" technology that underpins bitcoin, has emerged as an object of intense interest in the tech industry and beyond. Blockchain technology offers a way of recording transactions or any digital interaction in a way that is designed to be secure, transparent, highly resistant to outages, auditable and efficient; as such, it carries the possibility of disrupting industries and enabling new business models. The technology is young and changing very rapidly; widespread commercialization is still a few years off. Nonetheless, to avoid disruptive surprises or missed opportunities, strategists, planners and decision-makers across industries and business functions should pay heed now and begin to investigate applications of the technology [51].

11.5.1 What is Blockchain?

Blockchain is a database that maintains a continuously growing set of data records. It is distributed in nature, meaning that there is no master computer holding the entire chain. Rather, the participating nodes have a copy of the chain. It is also ever-growing – data records are only added to the chain [5].

A Blockchain consists of two types of elements:

- **Transactions** are the actions created by the participants in the system.
- **Blocks** record these transactions and ensure that they are in the correct sequence and have not been tampered with. Blocks also record a time stamp when the transactions were added.

11.5.2 What are Some Advantages of Blockchain?

There are three main advantages of Blockchain (Figure 11.1):

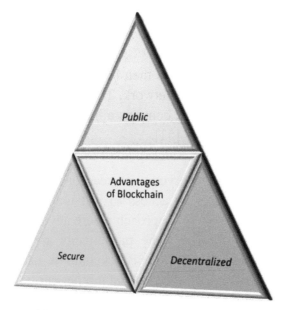

Figure 11.1 Advantages of Blockchain.

The big advantage of Blockchain is that it is *public*. Everyone participating can see the blocks and the transactions stored in them. This does not mean that everyone can see the actual content of your transaction; however, that is protected by your private key.

A Blockchain is *decentralized*, so there is no single authority that can approve the transactions or set specific rules to have transactions accepted. That means there is a huge amount of trust involved since all the participants in the network have to reach a consensus to accept transactions.

Most importantly, it is *secure*. The database can only be extended and previous records cannot be changed (at least, there is a very high cost if someone wants to alter previous records).

11.5.3 How Does It Work?

When someone wants to add a transaction to the chain, all the participants in the network will validate it. They do this by applying an algorithm to the transaction to verify its validity. What exactly is understood by "valid" is defined by the Blockchain system and can differ between systems. Then, it is up to a majority of the participants to agree that the transaction is valid.

A set of approved transactions is then bundled in a block, which is then sent to all the nodes in the network. They, in turn, validate the new block. Each successive block contains a hash, which is a unique fingerprint, of the previous block [51].

There are two main types of Blockchain:

- In a **public** Blockchain, everyone can read or write data. Some public Blockchains limit the access to just reading or writing. Bitcoin, for example, uses an approach where anyone can write.
- In a **private** Blockchain, all the participants are known and trusted. This is useful when the Blockchain is used between companies that belong to the same legal mother entity.

11.6 The Blockchain and IoT

Blockchain technology is the missing link to settle scalability, privacy and reliability concerns in the Internet of Things. Blockchain technologies could perhaps be the silver bullet needed by the IoT industry. Blockchain technology can be used in tracking billions of connected devices, enable the processing of transactions and coordination between devices and allow for significant savings to IoT industry manufacturers. This decentralized approach would eliminate single points of failure, creating a more resilient ecosystem for devices to run on. The cryptographic algorithms used by Blockchains would make consumer data more private.

The ledger is tamper-proof and cannot be manipulated by malicious actors because it does not exist in any single location, and man-in-the-middle attacks cannot be staged because there is no single thread of communication that can be intercepted. Blockchain makes trustless, peer-to-peer messaging possible and has already proven its worth in the world of financial services through cryptocurrencies such as Bitcoin, providing guaranteed peer-to-peer payment services without the need for third-party brokers.

The decentralized, autonomous and trustless capabilities of the Blockchain make it an ideal component to become a fundamental element of IoT solutions. It is not a surprise that enterprise IoT technologies have quickly become one of the early adopters of Blockchain technologies.

In an IoT network, the Blockchain can keep an immutable record of the history of smart devices. This feature enables the autonomous functioning of smart devices without the need for centralized authority. As a result, the Blockchain opens the door to a series of IoT scenarios that were remarkably difficult, or even impossible to implement without it.

By leveraging the Blockchain, IoT solutions can enable secure, trustless messaging between devices in an IoT network. In this model, the Blockchain will treat message exchanges between devices similar to financial transactions in a bitcoin network. To enable message exchanges, devices will leverage smart contracts, which then model the agreement between the two parties.

In this scenario, we can sensor from afar, communicating directly with the irrigation system in order to control the flow of water based on conditions detected on the crops. Similarly, smart devices in an oil platform can exchange data to adjust functioning based on weather conditions.

Using the Blockchain will enable true autonomous smart devices that can exchange data, or even execute financial transactions, without the need of a centralized broker. This type of autonomy is possible because the nodes in the Blockchain network will verify the validity of the transaction without relying on a centralized authority.

In this scenario, we can envision smart devices in a manufacturing plant that can place orders for repairing some of its parts without the need of human or centralized intervention. Similarly, smart vehicles in a truck fleet will be able to provide a complete report of the most important parts needing replacement after arriving at a workshop.

One of the most exciting capabilities of the Blockchain is the ability to maintain a duly decentralized, trusted ledger of all transactions occurring in a network. This capability is essential to enable the many compliances and regulatory requirements of industrial IoT applications without the need to rely on a centralized model [51, 52].

12

IoT and Blockchain: Challenges and Risks

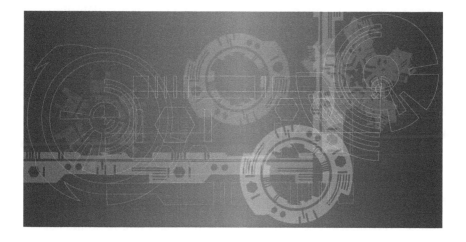

The Internet of Things (IoT) is an ecosystem of ever-increasing complexity; it is the next wave of innovation that will humanize every object in our life, and it is the next level of automation for every object we use. IoT is bringing more and more things into the digital fold every day, which will likely make IoT a multi-trillion dollar industry in the near future. To understand the scale of interest in the Internet of Things (IoT), just check how many conferences, articles and studies have been conducted about IoT recently. This interest has hit fever pitch point in 2016, as many companies see big opportunity and believe that IoT holds the promise to expand and improve business processes and accelerate growth. However, the rapid evolution of the IoT market has caused an explosion in the number and variety of IoT solutions, which created real challenges as the industry evolves, mainly the urgent need for a secure IoT model to perform common tasks such as sensing, processing, storage and communicating. Developing that model will never be an easy task by any stretch of the imagination, and there are many hurdles and challenges facing a real secure IoT model.

The biggest challenge facing IoT security is coming from the very architecture of the current IoT ecosystem; it is all based on a centralized model known as the server/client model. All devices are identified, authenticated and connected through cloud servers that support huge processing and storage capacities. The connection between devices will have to go through the cloud, even if they happen to be a few feet apart. While this model has connected computing devices for decades and will continue to support today IoT networks, it will not be able to respond to the growing needs of the huge IoT ecosystems of tomorrow.

12.1 The Blockchain Model

Blockchain is a database that maintains a continuously growing set of data records. It is distributed in nature, meaning that there is no master computer holding the entire chain. Rather, the participating nodes have a copy of the chain. It is also ever-growing – data records are only added to the chain.

When someone wants to add a transaction to the chain, all the participants in the network will validate it. They do this by applying an algorithm to the transaction to verify its validity. What exactly is understood by "valid" is defined by the Blockchain system and can differ between systems. Then, it is up to a majority of the participants to agree that the transaction is valid.

A set of approved transactions is then bundled in a block, which is sent to all the nodes in the network. They, in turn, validate the new block. Each successive block contains a hash, which is a unique fingerprint, of the previous block [53].

12.2 Principles of Blockchain Technology

Here are five basic principles underlying the technology [54].

1. Distributed Database

   Each party on a Blockchain has access to the entire database and its complete history. No single party controls the data or the information. Every party can verify the records of its transaction partners directly, without an intermediary.

2. Peer-to-Peer Transmission

   Communication occurs directly between peers instead of through a central node. Each node stores and forwards information to all other nodes.

3. Transparency

   Every transaction and its associated value are visible to anyone with access to the system. Each node, or user, on a Blockchain has a unique 30-plus-character alphanumeric address that identifies it. Users can choose to remain anonymous or provide proof of their identity to others. Transactions occur between Blockchain addresses.

4. Irreversibility of Records

   Once a transaction is entered in the database and the accounts are updated, the records cannot be altered, because they are linked to every transaction record that came before them (hence the term "chain"). Various computational algorithms and approaches are deployed to ensure that the recording on the database is permanent, chronologically ordered and available to all others on the network.

5. Computational Logic

   The digital nature of the ledger means that Blockchain transactions can be tied to computational logic and in essence programmed. Therefore, users can set up algorithms and rules that automatically trigger transactions between nodes.

12.3 Public vs. Private Blockchain

Blockchain technology implementation can be public or private with clear differences, for example, the benefits offered by a private Blockchain are: faster transaction verification and network communication, the ability to fix errors and reverse transactions and the ability to restrict access and reduce the likelihood of outsider attacks. The operators of a private Blockchain may choose to unilaterally deploy changes with which some users disagree. To ensure both the security and the utility of a private Blockchain system, operators must consider the recourse available to users who disagree with changes to the system's rules or are slow to adopt the new rules. However, developers who work to maintain public Blockchain systems like bitcoin still rely on individual users to adopt any changes they propose, which serves to ensure that changes are only adopted if they are in the interest of the entire system.

Just as a business will decide which of its systems are better hosted on a more secure private intranet or on the Internet, but will likely use both, systems requiring fast transactions, the possibility of transaction reversal, and central control over transaction verification will be better suited for private Blockchain, while those that benefit from widespread participation, transparency and third-party verification will flourish on a public Blockchain.

12.4 Challenges of Blockchain in IoT

In spite of all its benefits, the Blockchain model is not without flaws and shortcomings, (Figure 12.1) which are presented below [54]:

Scalability issues related to the size of Blockchain ledger that might lead to centralization as it is grown over time and require some kind of record management, which is casting a shadow over the future of the Blockchain technology.

Processing power and time required to perform encryption algorithms for all the objects involved in Blockchain-based IoT ecosystem given

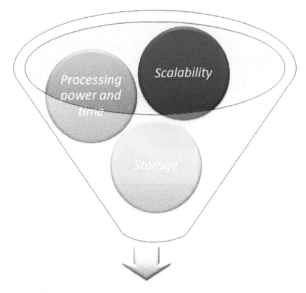

Challenges of Blockchain in IoT

Figure 12.1 Challenges of Blockchain in IoT.

the fact that IoT ecosystems are very diverse and composed of devices that have very different computing capabilities, and not all of them will be capable of running the same encryption algorithms at the desired speed.

Storage will be a hurdle: Blockchain eliminates the need for a central server to store transactions and device IDs, but the ledger has to be stored on the nodes themselves, and the ledger will increase in size as time passes. That is beyond the capabilities of a wide range of smart devices such as sensors, which have very low storage capacity.

12.5 Risks of Using Blockchain in IoT

It goes without saying that any new technology comes with new risks. An organization's risk management team should analyze, assess and design mitigation plans for risks expected to emerge from implementation of Blockchain-based frameworks (Figure 12.2).

Figure 12.2   Risks of using Blockchain in IoT.

Vendor Risks: Practically speaking, most present organizations, looking to deploy Blockchain-based applications, lack the required technical skills and expertise to design and deploy a Blockchain-based system and implement smart contracts completely in-house, i.e. without reaching out for vendors of Blockchain applications. The value of these applications is only as strong as the credibility of the vendors providing them. Given the fact that the Blockchain-as-a-Service (BaaS) market is still a developing market, a business should meticulously select a vendor that can perfectly sculpture applications that appropriately address the risks associated with the Blockchain.

Credential Security: Even though the Blockchain is known for its high security levels, a Blockchain-based system is only as secure as the system's access point. When considering a public Blockchain-based system, any individual who has access to the private key of a given user, which enables him/her to "sign" transactions on the public ledger, will effectively become that user, because most current systems do not provide multi-factor authentication. Also, loss of an account's private keys can lead to complete loss of funds, or data, controlled by this account; this risk should be thoroughly assessed.

Legal and Compliance: It is a new territory in all aspects without any legal or compliance precedents to follow, which poses a serious problem for IoT manufacturers and service providers. This challenge alone will scare off many businesses from using Blockchain technology.

12.6 The Optimum Secure IoT Model

In order for us to achieve that optimal secure model of IoT, security needs to be built in as the foundation of IoT ecosystem, with rigorous validity checks, authentication, data verification and all the data need to be encrypted at all levels, without a solid bottom–top structure, and we will create more threats with every device added to the IoT. What we need is a secure and safe IoT with privacy protected. That is a tough trade-off, but possible with Blockchain technology if we can overcome its drawbacks [55].

13

DDoS Attack: A Wake-Up Call for IoT

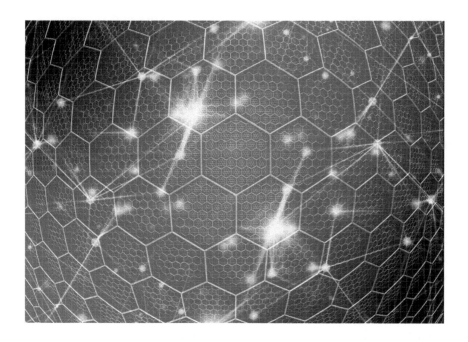

Welcome to the world of Internet of Things, wherein a glut of devices are connected to the Internet, which emanates massive amounts of data. Analysis and use of this data will have a real positive impact on our lives. But we have many hoops to jump before we can claim that crown, starting with a huge number of devices lacking unified platform with serious issues of security standards threatening the very progress of IoT.

The concept of IoT introduces a wide range of new security risks and challenges to IoT devices, platforms and operating systems, communications and even the systems to which they are connected. New security technologies will be required to protect IoT devices and platforms from both information attacks and physical tampering, to encrypt their communications and to address new challenges such as impersonating "things" or denial-of-sleep attacks that drain batteries, to denial-of-service attack (DoS). But IoT security will be complicated by the fact that many "things" use *simple processors* and *operating systems* that may not support sophisticated security approaches. In addition to all that, "Experienced IoT security specialists are scarce, and security solutions are currently fragmented and involve multiple vendors", said Mr. Jones from Gartner [58]. He added, "New threats will emerge through 2021 as hackers find new ways to attack IoT devices and protocols, so long-lived 'things' may need updatable hardware and software to adapt during their life span."

This fear was realized with a massive distributed denial of service attack that crippled the servers of services like Twitter, NetFlix, NYTimes and PayPal across the United States on October 21, 2016. It is the result of an immense assault that involved millions of Internet addresses and malicious software, according to Dyn, the prime victim of that attack. "One source of the traffic for the attacks was devices infected by the Mirai botnet" the link to the source code of Mirai malware on GitHub is here. The attack comes amid heightened cybersecurity fears and a rising number of Internet security breaches. Preliminary indications suggest that countless Internet of Things (IoT) devices that power everyday technology like closed-circuit cameras and smart-home devices were hijacked by the malware and used against the servers [60].

"Mirai scours the Web for IoT devices protected by little more than factory-default usernames and passwords, and then enlists the devices in attacks that hurl junk traffic at an online target until it can no longer accommodate legitimate visitors or users," Krebs explained in a post [61].

What makes this attack so interesting is that the devices hijacked have been networked to create the Internet of Things. In this case, the offender was likely digital video recorders, set-top boxes that allow you to record live TV and skip the commercials and webcams like those used around houses for security. All of these devices now moonlight as zombies under control of malicious actors bent on taking down individual websites or even portions of the Internet, as with the Dyn attack.

Considering the trend in connectivity, this is really just a taste of things to come. The deployment of IoT is far outpacing any other networked system. Gartner estimated that, by 2020, 20 billion devices will be connected to the Internet, that is, 20 billion new accomplices for an attacker to use to take down the servers that are critical to a functioning internet. Added to this explosion in connected (and potentially compromised) devices is the increasingly sophisticated and systematic nature of recent attacks.

13.1 Security is Not the Only Problem

A comprehensive study on IoT by the Internet Society (ISOC) revealed critical issues, which will have an impact on IoT [56]:

13.1.1 Security Concerns

With so many interconnected devices out there in market and plenty more to come in the near future, a security policy cannot be an afterthought, with the following issues with devices of IoT:

- Some devices are more secure than others
- Lack of updates on Internet of Things devices
- Communications security
- Consumer education

If the IoT devices are poorly secured, cyber attackers will use them as entry points to cause harm to other devices in the network. This will lead to loss of personal data out into the public, and the entire trust factor between Internet-connected devices and people using them will deteriorate.

In order to evade such scenarios, it is extremely critical to ensure the security, resilience and reliability of Internet applications to promote use of Internet-enabled devices among users across the world.

Security constraints for IoT are so critical that even analyst firm Gartner came out with some astounding numbers [57, 58].

- According to them, the worldwide spend for the IoT security market will reach US $348 million in 2016, a rise of 23.7% from US $281.5 million in 2015.
- Through 2018, over 50% of IoT device manufacturers will not be able to address threats from weak authentication practices.
- By 2020, more than 25% of identified enterprise attacks will involve IoT, although IoT will account for only 10% of IT security budgets.

13.1.2 Privacy Issues

The possibility of tracking and surveillance of people by government and private agencies increases as the devices are constantly connected to the Internet.

These devices collect user data without their permission, analyze them for purposes only known to the parent company. The social embrace of the IoT devices leads people to trust these devices with collection of their personal data without understanding the future implications.

13.1.3 Inter-Operatability Standard Issues

In an ideal environment, information exchange should take place between all the interconnected IoT devices. But the actual scenario is inherently more complex and depends on various levels of communication protocol stacks between such devices.

The OEMs producing industry-ready IoT devices will need to invest a lot of money and time to create standardized protocols common for all IoT devices or else it will delay product deployment across different verticals.

13.1.4 Legal Regulatory and Rights Issues

There are no concrete laws present which encompasses the various layers of IoT across the world. The array of devices connected to each other raises many security issues and no existing legal laws address such exposures.

The issues lie in whether current liability laws will extend their arm for devices that are connected to the Internet all the time because such devices have complex accountability issues.

13.1.5 Emerging Economy and Development Issues

IoT provides a great platform to enable social development in varied societies across the world, the proliferation of Internet across the various sections of the society in developing countries coupled with lowering costs of microprocessors and sensors will make IoT devices accessible to low-income households.

13.2 How to Prevent Future Attacks?

There are *four* interrelated things that need to change if we are to have a chance to combat this growing threat (Figure 13.1):

First, we need to change our culture around networked technologies, for example, not using default/generic passwords and disabling all remote (WAN) access to our devices.

Second, industry leaders need to make security and resilience in digital spaces a priority. When considering overall strategy, whether for an enterprise or a government, cyber strategy must be a key concern.

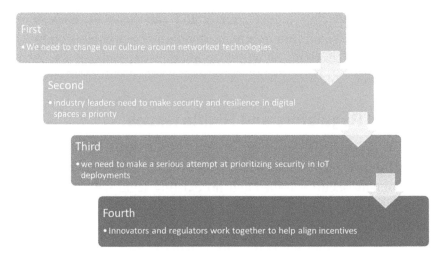

Figure 13.1   How to prevent future attacks.

Third, we need to make a serious attempt at prioritizing security in IoT deployments. Security by design, or ensuring that security is built into technology from the beginning, for example, security at the chip level is a step in the right direction.

Fourth, innovators and regulators work together to help align incentives, which are currently behind deploy-first-secure-later approaches, to support security in IoT.

PART IV

AI, Fog Computing and IoT

14

Why IoT Needs Fog Computing

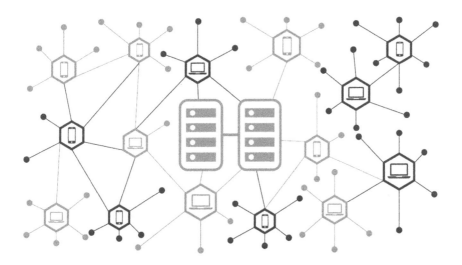

The Internet of Things (IoT) is one of the hottest mega-trends in technology, and for good reason, IoT deals with all the components of what we consider web 3.0, including Big Data Analytics, Cloud Computing and Mobile Computing.

14.1 The Challenge

The IoT promises to bring the connectivity to an earthly level, permeating every home, vehicle and workplace with smart, Internet-connected devices. But as dependence on our newly connected devices increases along with the benefits and uses of a maturing technology, the reliability of the gateways that make the IoT a functional reality must increase and make up time a near guarantee. As every appliance, light, door, piece of clothing and every other object in your home and office become potentially Internet-enabled, the Internet of Things is poised to apply major stresses to the current Internet and data center infrastructure.

The popular current approach is to centralize cloud data processing in a single site, resulting in lower costs and strong application security. But with the sheer amount of input data that will be received from globally distributed sources, this central processing structure will require backup. Also, most enterprise data is pushed up to the cloud, stored and analyzed, after which a decision is made and action taken. But this system is not efficient, and to make it efficient, there is a need to process some data or some Big Data in a smart way in the case of IoT, especially if it is sensitive data and needs quick action.

To illustrate the need for smart processing of some kind of data, IDC estimates that the amount of data analyzed on devices that are physically close to the Internet of Things is approaching 40%, which supports the urgent need for a different approach to this need [63].

14.2 The Solution

To deal with this challenge, *fog computing* is the champion.

Fog computing allows computing, decision-making and action-taking to happen via IoT devices and only pushes relevant data to the cloud. Cisco coined the term "fog computing" and gave a brilliant definition for it: "The fog extends the cloud to be closer to the things that produce and act on IoT data. These devices, called fog nodes, can be deployed anywhere with a network connection: on a factory floor, on top of a power pole, alongside a railway track, in a vehicle, or on an oil rig. Any device with computing, storage, and network connectivity can be a fog node. Examples include industrial controllers, switches, routers, embedded servers, and video surveillance cameras." [62]

To understand the concept of fog computing, the following actions define fog computing:

- Analyzes the most time-sensitive data at the network edge, close to where it is generated instead of sending vast amounts of IoT data to the cloud.
- Acts on IoT data in milliseconds, based on policy.
- Sends selected data to the cloud for historical analysis and longer-term storage.

14.3 Benefits of Using Fog Computing

- Minimize latency
- Conserve network bandwidth
- Address security concerns at all levels of the network
- Operate reliably with quick decisions
- Collect and secure wide range of data
- Move data to the best place for processing
- Lower expenses of using high computing power only when needed and less bandwidth
- Better analysis and insights of local data

Keep in mind that fog computing is not a replacement of cloud computing by any measure; it works in conjunction with cloud computing, optimizing the use of available resources. But it was the product of a need to address two challenges, real-time process and action of incoming data, and limitation of resources like bandwidth and computing power, another factor helping fog computing is the fact that it takes advantage of the distributed nature of today's virtualized IT resources. This improvement to the data-path hierarchy is enabled by the increased compute functionality that manufacturers are building into their edge routers and switches.

14.4 Real-Life Example

A traffic light system in a major city is equipped with smart sensors. It is the day after the local team won a championship game, and it is the morning of the day of the big parade. A surge of traffic into the city is expected as revelers come to celebrate their team's win. As the traffic builds, data are collected from individual traffic lights. The application developed by the city to adjust light patterns and timing is running on each edge device. The app automatically makes adjustments to light patterns in real time, at the edge, working around traffic impediments as they arise and diminish. Traffic delays are kept to a minimum, and fans spend less time in their cars and have more time to enjoy their Big Day.

After the parade is over, all the data collected from the traffic light system would be sent up to the cloud and analyzed, supporting predictive analysis and allowing the city to adjust and improve its traffic application's response to future traffic anomalies. There is little value in sending a live steady stream of everyday traffic sensor data to the cloud for storage and analysis. The civic engineers have a good handle on normal traffic patterns. The relevant data is sensor information that diverges from the norm, such as the data from parade day [64].

14.5 The Dynamics of Fog Computing

Fog computing thought of as a "low to the ground" extension of the cloud to nearby gateways and proficiently provides for this need. As Gartner's networking analyst, Joe Skorupa puts it: "The enormous number of devices, coupled with the sheer volume, velocity, and structure of IoT data, creates challenges, particularly in the areas of security, data, storage management, servers and the data center network with real-time business processes at stake. Data center managers will need to deploy more forward-looking capacity management in these areas to be able to proactively meet the business priorities associated with IoT."

For data handling and backhaul issues that shadow the IoT's future, fog computing offers a functional solution. Networking equipment vendors proposing such a framework envision the use of routers with industrial-strength reliability, running a combination of open Linux and JVM platforms embedded with vendor's own proprietary OS. By using open platforms, applications could be ported to IT infrastructure using a programming environment that is familiar and supported by multiple vendors. In this way, smart edge gateways can either handle or intelligently redirect the millions of tasks coming from the myriad sensors and monitors of the IoT, transmitting only summary and exception data to the cloud [62].

14.6 Fog Computing and Smart Gateways

The success of fog computing hinges directly on the resilience of those smart gateways directing countless tasks on the Internet teeming with IoT devices. IT resilience will be a necessity for the business continuity of IoT operations, with the following tasks to ensure that success:

- Redundancy
- Security
- Monitoring of power and cooling
- Failover solutions in place to ensure maximum uptime

According to Gartner [114], every hour of downtime can cost an organization up to US $300,000. The speed of deployment, cost-effective scalability and ease of management with limited resources are also chief concerns.

14.7 Conclusion

Moving the intelligent processing of data to the edge only raises the stakes for maintaining the availability of these smart gateways and their communication path to the cloud. When the IoT provides methods that allow people to manage their daily lives, from locking their homes to checking their schedules to cooking their meals, gateway downtime in the fog computing world becomes a critical issue. Additionally, resilience and failover solutions that safeguard those processes will become even more essential.

15

AI Is the Catalyst of IoT

Businesses across the world are rapidly leveraging the Internet of Things (IoT) to create new products and services that are opening up new business opportunities and creating new business models. The resulting transformation is ushering in a new era of how companies run their operations and engage with customers. However, tapping into the IoT is only part of the story [70].

For companies to realize the full potential of IoT enablement, they need to combine IoT with rapidly advancing artificial intelligence (AI) technologies, which enable "smart machines" to simulate intelligent behavior and make well-informed decisions with little or no human intervention [70].

Artificial intelligence (AI) and Internet of Things (IoT) are terms that project futuristic, sci-fi imagery; both have been identified as drivers of business disruption in 2017. But, what do these terms really mean and what is their relation? Let us start by defining both terms first:

IoT is defined as a system of interrelated *physical objects, sensors, actuators, virtual objects, people, services, platforms and networks* that have separate identifiers and an ability to transfer data independently. Practical examples of IoT application today include precision agriculture, remote patient monitoring and driverless cars. Simply put, IoT is the network of "things" that collects and exchanges information from the environment [71].

IoT is sometimes referred to as the driver of the *Fourth Industrial Revolution* (Industry 4.0) [72] by industry insiders and has triggered technological changes that span a wide range of fields. Gartner forecasted that there would be 20 billion connected things in use worldwide by 2020, but more recent predictions put the 2020 figure at over 50 billion devices [68] .Various other reports have predicted huge growth in a variety of industries, such as estimating healthcare IoT to be worth US \$117 billion by 2020 and forecasting 250 million connected vehicles on the road by the same year. IoT developments bring exciting opportunities to make our personal lives easier as well as improving efficiency, productivity and safety for many businesses [66].

AI, on the contrary, is the engine or the "brain" that will enable analytics and decision-making from the data collected by IoT. In other words, IoT collects the data and AI processes this data in order to make sense of it. You can see these systems working together at a personal level in devices like fitness trackers and Google Home, Amazon's Alexa and Apple's Siri [65].

With more connected devices comes more data that has the potential to provide amazing insights for businesses but presents a new challenge for how to analyze it all. Collecting this data benefits no one unless there is a way to understand it all. This is where AI comes in. Making sense of huge amounts of data is a perfect application for pure AI.

By applying the analytic capabilities of AI to data collected by IoT, companies can identify and understand patterns and make more informed decisions. This leads to a variety of benefits for both consumers and companies such as proactive intervention, intelligent automation and highly personalized experiences. It also enables us to find ways for connected devices to work better together and make these systems easier to use.

This, in turn, leads to even higher adoption rates. That is exactly why we need to improve the speed and accuracy of data analysis with AI in order to see IoT live up to its promise. Collecting data is one thing, but sorting, analyzing and making sense of that data is a completely different thing. That is why it is essential to develop faster and more accurate AIs in order to keep up with the sheer volume of data being collected as IoT starts to penetrate almost all aspects of our lives.

15.1 Examples of IoT Data [68]

- Data that helps cities predict accidents and crimes
- Data that gives doctors real-time insight into information from pacemakers or biochips
- Data that optimizes productivity across industries through predictive maintenance on equipment and machinery
- Data that creates truly smart homes with connected appliances
- Data that provides critical communication between self-driving cars

It is simply impossible for humans to review and understand all of this data with traditional methods, even if they cut down the sample size, it simply takes too much time. The big problem will be finding ways to analyze the deluge of performance data and information that all these devices create. Finding insights in terabytes of machine data is a real challenge, just ask a data scientist.

But in order for us to harvest the full benefits of IoT data, we need to improve:

- Speed of Big Data analysis
- Accuracy of Big Data analysis

15.2 AI and IoT Data Analysis

There are six types of IoT data analysis where AI can help [69]:

1. *Data preparation*: Defining pools of data and clean them, which will take us to concepts like Dark Data and Data Lakes.
2. *Data discovery*: Finding useful data in the defined pools of data
3. *Visualization of streaming data*: On the fly dealing with streaming data by defining, discovering data and visualizing it in smart ways to make it easy for the decision-making process to take place without delay.
4. *Time series accuracy of data*: Keeping the level of confidence in data collected high with high accuracy and integrity of data
5. *Predictive and advance analysis*: A very important step where decisions can be made based on data collected, discovered and analyzed.
6. *Real-time geospatial and location (logistical data)*: Maintaining the flow of data smooth and under control.

15.3 AI in IoT Applications

- Visual Big Data, for example, will allow computers to gain a deeper understanding of images on the screen, with new AI applications that understand the context of images.
- Cognitive systems will create new recipes that appeal to the user's sense of taste, creating optimized menus for each individual, and automatically adapting to local ingredients.
- Newer sensors will allow computers to "hear" gathering sonic information about the user's environment.

- Connected and remote operations: With smart and connected warehouse operations, workers no longer have to roam the warehouse picking goods off the shelves to fulfill an order. Instead, shelves whisk down the aisles, guided by small robotic platforms that deliver the right inventory to the right place, avoiding collisions along the way. Order fulfillment is faster, safer and more efficient.
- Prevented/predictive maintenance: Saving companies millions before any breakdown or leaks by predicting and preventing locations and time of such events.

These are just a few promising applications of artificial intelligence in IoT. The potential for highly individualized services are endless and will dramatically change the way people live.

15.4 Challenges Facing AI in IoT

Challenges facing AI in IoT can be summarized in the following points (Figure 15.1) [68, 73]:

1. Compatibility: IoT is a collection of many parts and systems, which are fundamentally different in time and space.
2. Complexity: IoT is a complicated system with many moving parts and nonstop stream of data, making it a very complicated ecosystem.
3. Privacy/security/safety (PSS): PSS is always an issue with every new technology or concept, how far IA can help without compromising PSS? One of the new solutions for such problem is using Blockchain technology.
4. Ethical and legal issues: It is a new world for many companies with no precedents, untested territory with new laws and cases emerging rapidly.
5. Artificial stupidity: Back to the very simple concept of GIGO (garbage in garbage out), AI still needs "training" to understand human reactions/emotions so the decisions will make sense.

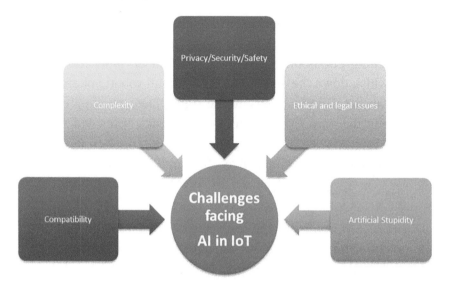

Figure 15.1   Challenges facing AI in IoT.

15.5 Conclusion

While IoT is quite impressive, it really does not amount to much without a good AI system. Both technologies need to reach the same level of development in order to function as perfectly as we believe they should and would. Scientists are trying to find ways to make more intelligent data analysis software and devices in order to make safe and effective IoT a reality. It may take some time before this happens because AI development is lagging behind IoT, but the possibility is, nevertheless, there.

Integrating AI into IoT networks is becoming a prerequisite for success in today's IoT-based digital ecosystems. Therefore, businesses must move rapidly to identify how they will drive value from combining AI and IoT – or face playing catch-up in years to come.

The only way to keep up with this IoT-generated data and gain the hidden insights it holds is using AI as the catalyst of IoT.

PART V

The Future of IoT

16

IoT, AI and Blockchain: Catalysts for Digital Transformation

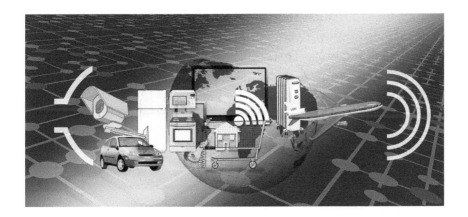

The digital revolution has brought with it a new way of thinking about manufacturing and operations. Emerging challenges associated with logistics and energy costs are influencing global production and associated distribution decisions. Significant advances in technology, including Big Data analytics, AI, Internet of Things, robotics and additive manufacturing, are shifting the capabilities and value proposition of global manufacturing. In response, manufacturing and operations require a digital renovation: the value chain must be redesigned and retooled and the workforce retrained. Total delivered cost must be analyzed to determine the best places to locate sources of supply, manufacturing and assembly operations around the world. In other words, we need a digital transformation.

16.1 Digital Transformation

Digital transformation (DX) is the profound transformation of business and organizational activities, processes, competencies and models to fully leverage the changes and opportunities of a mix of digital technologies and their accelerating impact across society in a strategic and prioritized way, with present and future shifts in mind (Figure 16.1).

A digital transformation strategy aims to create the capabilities of fully leveraging the possibilities and opportunities of new technologies and their impact faster, better and in more innovative way in the future.

A digital transformation journey needs a staged approach with a clear road-map, involving a variety of stakeholders, beyond silos and internal/external limitations. This road-map takes into account that end goals will continue to move as digital transformation *de facto* is an ongoing journey, as is change and digital innovation [75].

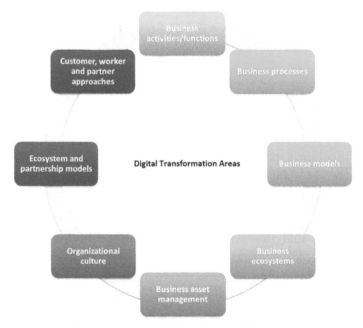

Figure 16.1   Digital Transformation Areas.

16.2 Internet of Things (IoT)

IoT is defined as a system of interrelated *physical objects, sensors, actuators, virtual objects, people, services, platforms and networks* that have separate identifiers and an ability to transfer data independently. Practical examples of IoT application today include precision agriculture, remote patient monitoring and driverless cars. Simply put, IoT is the network of "things" that collects and exchanges information from the environment.

IoT and digital transformation are closely related for the following reasons [115,116]:

1. More than 50% of companies think IoT is strategic, and one in four believes it is transformational.
2. Both increase company longevity. The average life span of a company has decreased from 67 years in the 1920s to 15 years today.
3. One in three industry leaders will be digitally disrupted by 2018.
4. Both enable businesses to connect with customers and partners in open digital ecosystems, to share digital insights, collaborate on solutions and share in the value created.
5. Competitors are doing it. According to IDC, 70% of global discrete manufacturers will offer connected products by 2019.
6. It is where the money is. Digital product and service sales are growing and will represent more than US $1 of every US $3 spent by 2021.
7. Enterprises are overwhelmed by data and digital assets. They already struggle to manage the data and digital assets they have, and IoT will expand them exponentially. They need help finding the insights in the vast stream of data and manage digital assets.
8. Both drive consumption. Digital services easily prove their own worth. Bundle products with digital services and content make it easy for customers to consume them.
9. Both make companies understand customers better. Use integrated channels, Big Data, predictive analytics and machine

learning to uncover, predict and meet customer needs, increasing loyalty and revenues, IoT and AI are at the heart of this.

10. Using both is future-proof for the business. Make the right strategic bets for the company, product and service portfolio and future investments using IoT data analytics, visualization and AI.

16.3 Digital Transformation, Blockchain and AI

Digital transformation is a complicated challenge, but the integration of Blockchain and AI makes it much easier. Considering the number of partners (internal, external or both) involved in any given business process, a system in which a multitude of electronic parties can securely communicate, collaborate and transact without human intervention is highly agile and efficient.

Enterprises that embrace this transformation will be able to provide a better user experience, a more consistent workflow, more streamlined operations and value-added services, as well as gain competitive advantage and differentiation.

Blockchain can holistically manage steps and relationships where participants will share the same data source, such as financial relationships and transactions connected to each step, security and accountability factored in, as well as compliance with government regulations along with internal rules and processes. The result is consistency, reductions in costs and time delays, improved quality and reduced risks [74].

AI can help companies learn in ways that accelerate innovation and assist companies getting closer to customers and improve employee's productivity and engagement. Digital transformation efforts can be improved with that information.

16.4 Conclusion

The building blocks of digital transformation are mindset, people, process and tools. IoT covers all the blocks since IoT does not just connect devices, it connects people too. Blockchain will ensure end-to-end security, and by using AI, you will move IoT beyond connections to intelligence. One important step is to team up with the best partners and invest in education, training and certifying your teams. This magical mix of IoT, AI and Blockchain will help make transformation digital and easy [77].

17

Future Trends of IoT

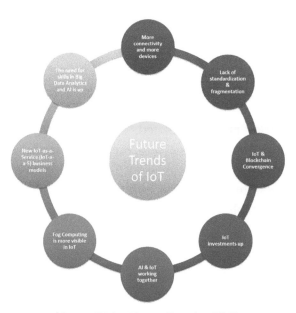

Figure 17.1 Future Trends of IoT.

IoT technology continues to evolve at an incredibly rapid pace. Consumers and businesses alike are anticipating the next big innovation. They are all set to embrace the ground-breaking impact of the Internet of Things on our lives like ATMs that report crimes around them, forks that tell you if you are eating fast or IP address for each organ of your body for doctors to connect and check [82].

IoT will see tremendous growth in all directions. The following eight trends (Figure 17.1) are the main developments we predict for coming years.

17.1 Trend 1 – Lack of Standardization Will Continue

Digitally connected devices are rapidly becoming an essential part of our everyday lives. Although the adoption of IoT will be large, it will most likely be slow. The primary reason for this is lack of standardization.

Although industry leaders are trying to develop specified standards and get rid of fragmentation, it will still exist. There will be no clear standards in the near future of IoT. Unless a well-respected organization like IEEE stepped-in and leads the way or the government imposes restrictions on doing business with companies if they are not using unified standards [83].

The hurdles facing IoT standardization can be divided into four categories: platform, connectivity, business model and killer applications.

All four categories are inter-related, and you need all them to make all them work. Missing one will break that model and stall the standardization process. A lot of work is needed in this process, and many companies are involved in each one of the categories, and bringing them to the table to agree on a unifying standard will be a daunting task [89].

17.2 Trend 2 – More Connectivity and More Devices

The speedy proliferation of IoT in past 3 years has resulted in billions of interconnected devices. As the consumer continues to stay hooked to more gadgets, the number of connected devices grew exponentially every year. By 2018, it will at least double and touch a whopping the mark of 46 billion by 2021. More IoT devices will enter the channels, more than ever before, a clear indication of our direct dependency over the gadgets and that is how our future is shaped [83].

As IoT continues to expand, we will certainly see an increase in devices connected to the network in different areas in business and consumer markets. Smart devices will become the *de facto* for people to manage IoT devices. The benefits of using smart devices in that capacity include boosting customer engagement, increasing visibility and streamlining communication that will include new human–machine interfaces such as voice user interface (VUI) or Chatbot [79, 82].

17.3 Trend 3 – "New Hope" for Security: IoT & Blockchain Convergence

As with most technology, security will be the major challenge that needs to be addressed. As the world becomes increasingly high-tech, devices are easily targeted by cyber-criminals. Evans Data states that 92% of IoT developers say that security will continue to be an issue in the future. Consumers not only have to worry about smartphones, but also other devices such as baby monitors, cars with Wi-Fi, wearables and medical devices can be breached. Security undoubtedly is a major concern, and vulnerabilities need to be addressed.

Blockchain is a "new hope" for IoT security. The astounding conquest of cryptocurrency, which is built on Blockchain technology, has put the technology as the flag bearer of seamless transactions, thereby reducing costs and doing away with the need to trust a centered data source.

Blockchain works by enhancing trustful engagements in a secured, accelerated and transparent pattern of transactions. The real-time data from an IoT channel can be utilized in such transactions while preserving the privacy of all parties involved [79, 82].

The big advantage of Blockchain is that it is *public*. Everyone participating can see the blocks and the transactions stored in them. This does not mean that everyone can see the actual contents of your transaction, however; that is protected by your private key.

A Blockchain is *decentralized*, so there is no single authority that can approve the transactions or set specific rules to have transactions accepted. That means there is a huge amount of trust involved since all the participants in the network have to reach a consensus to accept transactions.

Most importantly, it is *secure*. The database can only be extended and previous records cannot be changed (at least, there is a very high cost if someone wants to alter previous records) [80, 81, 84, 87]

In coming years, increased interest in Blockchain technology will make the convergence of Blockchain and IoT devices and services the next logical step for manufacturers and vendors, and many will compete for labels like "Blockchain Certified".

17.4 Trend 4 – IoT Investments Will Continue

IDC predicts that spending on IoT will reach nearly US $1.4 trillion in 2021. This coincides with companies continuing to invest in IoT hardware, software, services and connectivity. Almost every industry will be affected by IoT, which means many companies will benefit from its rapid growth. The largest spending category until 2021 will be hardware, especially modules and sensors, but is expected to be overtaken by the faster growing services category. Software spending will be similarly dominated by applications software including mobile apps.

IoT's undeniable impact will continue to lure more startup venture capitalists toward highly innovative projects. It is one of those few markets that have the interest of the emerging as well as traditional venture capital. While the growth next year is firmly attested and the true potential is yet to be unearthed, IoT ventures will be preferred over everybody else. Many businesses have assured adding IoT to their services model from the transportation, retail, insurance and mining industries [81, 83].

17.5 Trend 5 – Fog Computing Will Be More Visible

Fog computing allows computing, decision-making and action-taking to happen via IoT devices and only pushes relevant data to the cloud. Cisco coined the term "fog computing" and gave a brilliant definition for it: "The fog extends the cloud to be closer to the things that produce and act on IoT data. These devices, called *fog nodes*, can be deployed anywhere with a network connection: on a factory floor, on top of a power pole, alongside a railway track, in a vehicle, or on an oil rig. Any device with *computing, storage,* and *network connectivity* can be a fog node. Examples include industrial controllers, switches, routers, embedded servers, and video surveillance cameras."

The benefits of using fog computing are very attractive to IoT solution providers, some of which are minimizing latency, conserving network bandwidth, operating reliably with quick decisions, collecting and securing a wide range of data and moving data to the best place for processing with better analysis and insights of local data. Blockchain can be implemented at the level of fog nodes too [88].

17.6 Trend 6 – AI & IoT Will Work Closely

Amalgamation of IoT data analytics with AI for applications ranging from elevator maintenance to smart homes will progress rapidly over the coming 2 years. Platform and service providers are increasingly delivering solutions with integrated analytics designed to feed data directly into AI algorithms. Another important advantage of using AI is supporting the optimization and adaptation of both IoT devices and related processes and infrastructure.

AI can help IoT Data Analysis in the following areas: *data preparation, data discovery, visualization of streaming data, time series accuracy of data, predictive and advance analytics and real-time geospatial and location (logistical data)* [87].

17.7 Trend 7 – New IoT-as-a-Service (IoT-a-a-S) Business Models

Transformational business models will develop in many IoT verticals over 2018–2019, supported by Big Data and AI tools. In these models, the value is in the *convenience of the service* for end customers (on-demand and not requiring heavy upfront expenditure), and the *usage data* that is collected, analyzed and fed back into suppliers' businesses and processes.

But the potential for IoT business model transformation extends beyond this, to encompass an increasing variety of more complex, as-a-service business models that disrupt existing industries, particularly for areas such as heavy industry, transport and logistics and smart cities.

For these industries, IoT solutions can enable more of an ongoing, managed service relationship with both technology providers and end customers. One selling point is that costs can be more directly linked to ongoing measured usage or to specific trigger events captured by IoT sensors (e.g., "break-the-glass" solutions in which sensors pick up when a building or car is broken into). Another is that costs may be spread over time, shifting from upfront Capex to a more regular Opex outflow. Examples of such models include lighting-as-a-service (L-a-a-S), rail-as-a-service (R-a-a-S) and even elevators-as-a-service (E-a-a-S) [78].

17.8 Trend 8 – The Need for Skills in IoT's Big Data Analysis and AI Will Increase

Dynamic data sharing is at the heart of IoT and Big Data analysis will be instrumental in building responsive applications. Integrating IoT data channels with AI to retrieve on demand analytical insights has already gained momentum in 2017 and was expected to definitely grow exponentially in 2018 and beyond. Subsequently, the need for Big Data and AI skills will rise, while most IoT service providers have highlighted the shortage for such extensively skilled candidates, internal learning programs in close proximity with R&D has set to be launched in many companies [78, 85, 87].

PART VI

Inside Look at Blockchain

18

Myths about Blockchain Technology

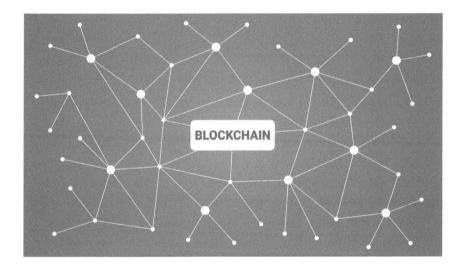

Blockchain, the "distributed ledger" technology, has emerged as an object of intense interest in the tech industry and beyond. Blockchain technology offers a way of recording transactions or any digital interaction in a way that is designed to be secure, transparent, highly resistant to outages, auditable and efficient; as such, it carries the possibility of disrupting industries and enabling new business models. The technology is young and changing very rapidly; widespread commercialization is still a few years off. Nonetheless, to avoid disruptive surprises or missed opportunities, strategists, planners and decision-makers across industries and business functions should pay heed now and begin to investigate applications of the technology.

Blockchain is a database that maintains a continuously growing set of data records. It is distributed in nature, meaning that there is no master computer holding the entire chain. Rather, the participating nodes have a copy of the chain. It is also ever-growing – data records are only added to the chain.

A Blockchain consists of two types of elements [97]:

- Transactions are the actions created by the participants in the system.
- Blocks record these transactions and make sure they are in the correct sequence and have not been tampered with.

The big advantage of Blockchain is that it is public. Everyone participating can see the blocks and the transactions stored in them. This does not mean everyone can see the actual content of your transaction, however; that is protected by your private key.

A Blockchain is decentralized, so there is no single authority that can approve the transactions or set specific rules to have transactions accepted. This means that there is a huge amount of trust involved since all the participants in the network have to reach a consensus to accept transactions.

Most importantly, it is secure. The database can only be extended and previous records cannot be changed (at least, there is a very high cost if someone wants to alter previous records).

When someone wants to add a transaction to the chain, all the participants in the network will validate it. They do this by applying an algorithm to the transaction to verify its validity. What exactly is understood by "valid" is defined by the Blockchain system and can differ between systems. Then, it is up to a majority of the participants to agree that the transaction is valid.

A set of approved transactions is then bundled in a block, which is then sent to all the nodes in the network. They, in turn, validate the new block. Each successive block contains a hash, which is a unique fingerprint, of the previous block.

Blockchain ensures that data has not been tampered with, offering a layer of time-stamping that removes multiple levels of human checking and makes transactions immutable. However, it is not yet the cure-all that some believe it to be [97].

There are three types of Blockchains (Figure 18.1) [96]:

- Public: a public Blockchain is the one where everyone can see all the transactions, anyone can expect their transaction to appear on the ledger and finally anyone can participate in the consensus process.
- Federated: federated Blockchain do not allow everyone to participate to the consensus process. Indeed, only a limited number of nodes are given the permission to do so. For instance, in a group of 20 pharmaceutical companies, we could imagine that for a block to be valid, 15 of them have to agree. The access to the Blockchain, however, can be public or restricted to the participants.

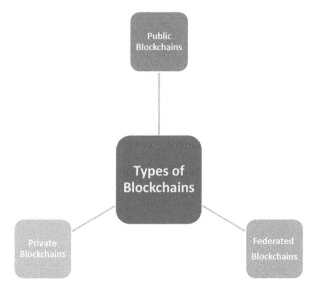

Figure 18.1   Types of Blockchains.

- Private: private Blockchains are generally used inside a company. Only specific members are allowed to access it and carry out transactions.

Blockchain technology certainly has many positive aspects, but there is also much misunderstanding and confusion regarding its nature.

18.1 Myth 1: The Blockchain is a Magical Database in the Cloud [95]

The Blockchain is conceptually a flat file – a linear list of simple transaction records. "This list is appended only so entries are never deleted, but instead, the file grows indefinitely and must be replicated in every node in the peer-to-peer network."

Blockchain does not allow you to store any type of physical information like a Word document or a pdf file. It can only provide a "proof-of-existence", and the distributed ledger can only contain a code that certifies the existence of a certain document but not the document itself. The file, however, can be stored in "data lakes", the access to which is controlled by the owner of the information.

18.2 Myth 2: Blockchain is Going to Change the World [94]

We can use Blockchain for complex and technical transactions such as verifying the authenticity of a diamond or the identity of a person. There is also talk of a Blockchain application for the bill of lading in trade finance, which would be revolutionary in terms of cost reduction and transaction speed.

While Blockchain can support these cases and mitigate the risk of a fraudster tampering with the ledger, it does not eradicate the threat of fraud online and it still raises questions over confidentiality. Additionally, the use of Blockchain technology will still be inefficient for many of these cases when compared to maintaining a traditional ledger.

18.3 Myth 3: Blockchain is Free [94]

Despite the commonly held belief, Blockchain is neither cheap nor efficient to run. However, it involves multiple computers solving mathematical algorithms to agree a final immutable result, which becomes the so-called single version of truth (SVT). Each "block" in the Blockchain typically uses a large amount of computing power to solve. And someone needs to pay for all this computer power that supports the Blockchain service.

18.4 Myth 4: There is Only One Blockchain [94]

There are many different technologies that go by the name Blockchain. They come in public and private versions, open and closed source, general purpose and tailored to specific solutions.

Common denominator is that they are shore up by crypto, are distributed and have some form of consensus mechanism. Bitcoin's Blockchain, Ethereum, Hyperledger, Corda, and IBM and Microsoft's Blockchain-as-a-Service can all be classified as distributed ledger technologies.

18.5 Myth 5: The Blockchain Can Be Used for Anything and Everything [93]

Although the code is powerful, it is not magical. Bitcoin and Blockchain developers can be evangelical, and it is easy to understand why. For many, the Blockchain is an authority tied to mathematics, not the government or lawyers. In the minds of some developers, the Blockchain and smart contracts will one day replace money, lawyers and other arbitration bodies. Yet the code is limited to the number of cryptocurrency transactions in the chain itself, and cryptocurrency is still far from mainstream.

18.6 Myth 6: The Blockchain Can Be the Backbone of a Global Economy [93]

No national or corporate entity owns or controls the Blockchain. For this reason, evangelists hope that private Blockchains can provide foundational support for dozens of encrypted and trusted cryptocurrencies. Superficially, the Bitcoin Blockchain appears massive. Yet a Gartner report has recently claimed that the size of the Blockchain is similar in scale to the NASDAQ network. If cryptocurrency takes off, and records are generated larger, this may change. For now, the Blockchain network is roughly analogous to contemporary financial networks.

18.7 Myth 7: The Blockchain Ledger is Locked and Irrevocable

Analogous large-scale transaction databases like bank records are, by their nature, private and tied to specific financial institutions. The power of Blockchain, of course, is that the code is public, transactions are verifiable and the network is cryptographically secure. Fraudulent transactions – double spends, in industry parlance – are rejected by the network, preventing fraud. Because mining the chain provides financial incentive in the form of Bitcoin, it is largely believed that rewriting historic transactions is not in the financial interest of participants. For now, however, as computational resources improve with time, so too does the potential for deception. The impact of future processing power on the integrity of the contemporary Blockchain remains unclear.

18.8 Myth 8: Blockchain Records Can Never Be Hacked or Altered [92]

One of the main selling points about Blockchains is their inherent permanence and transparency. When people hear that, they often think that Blockchains are invulnerable to outside attacks. No system or database will ever be completely secure, but the larger and more distributed the network, the more secure it is believed to be. What Blockchains can provide to applications that are developed on top of them is a way of catching unauthorized changes to records.

18.9 Myth 9: Blockchain Can Only Be Used in the Financial Sector [91]

Blockchain started to create waves in the financial sector because of its first application, the bitcoin cryptocurrency, which directly impacted this field. Although Blockchain has numerous areas of application, finance is undeniably one of them. The important challenges that this technology brings to the financial world pushed international banks such as Goldman Sachs or Barclays to heavily invest in it. Outside the financial sector, Blockchain can and will be used in real estate, healthcare or even at a personal scale to create a digital identity. Individuals could potentially store a proof-of-existence of medical data on the Blockchain and provide access to pharmaceutical companies in exchange for money.

18.10 Myth 10: Blockchain is Bitcoin [91]

Since Bitcoin is more famous than the underlying technology, Blockchain, many people get confused between the two.

Blockchain is a technology that allows peer-to-peer transactions to be recorded on a distributed ledger across the network. These transactions are stored in blocks and each block is linked to the previous one, therefore creating a chain. Thus, each block contains a complete and time-stamped record of all the transactions that occurred in the network. On the Blockchain, everything is transparent and permanent. No one can change or remove a transaction from the ledger.

Bitcoin is a cryptocurrency that makes electronic payment possible directly between two people without going through a third party like a bank. Bitcoins are created and stored in a virtual wallet. Since there are no intermediaries between the two parties, no one can control the cryptocurrency. Hence, the number of bitcoins that will ever be released is limited and defined by a mathematical algorithm.

18.11 Myth 11: Blockchain is Designed for Business Interactions Only [91]

Experts in Blockchain are convinced that this technology will change the world and the global economy just like dot-coms did in the early 1990s. Hence, it is not only open to big corporations, but is also accessible to everyone everywhere. If all it takes is an Internet connection to use the Blockchain, one can easily imagine how many people worldwide will be able to interact with each other.

18.12 Myth 12: Smart Contracts Have the Same Legal Value as Regular Contracts

For now, smart contracts are just pieces of code that execute actions automatically when certain conditions are met. Therefore, they are not considered as regular contracts from a legal perspective. However, they can be used as a proof of whether or not a certain task has been accomplished. Despite their uncertain legal value, smart contracts are very powerful tools especially when combined with the Internet of Things (IoT).

19

Cybersecurity & Blockchain

With the fact that cybercrime and cyber security attacks hardly seem to be out of the news these days and the threat is growing globally, nobody would appear immune to malicious and offensive acts targeting computer networks, infrastructures and personal computer devices. Firms must clearly invest to stay resilient. Gauging the exact size of cybercrime and putting a precise US dollar value on it is nonetheless tricky. But one thing we can be sure about is that the number is big and probably larger than the statistics reveal.

The global figure for cyber breaches had been put at around US $200 billion annually [117]. Malicious cyber activity cost the U.S. economy between $57 billion and $109 billion in 2016, the White House Council of Economic Advisers estimated in a report released in February of 2018 [118].

New Blockchain platforms are stepping up to address security concerns in the face of recent breaches. Since these platforms are not controlled by a singular entity, they can help ease the concerns created by a spree of recent breach disclosures. Services built on top of Blockchain have the potential to inspire renewed trust due to the transparency built into the technology.

Developments in Blockchain have expanded beyond recordkeeping and cryptocurrencies. The integration of *smart contract* development in Blockchain platforms has ushered in a wider set of applications, including cybersecurity.

By using Blockchain, transaction details are kept both transparent and secure. Blockchain's decentralized and distributed network also helps businesses to avoid a single point of failure, making it difficult for malicious parties to steal or tamper with business data.

Transactions in the Blockchain can be audited and traced. In addition, public Blockchains rely on distributed network to run, thus eliminating a single point of control. For attackers, it is much more difficult to attack a large number of peers distributed globally as opposed to a centralized data center.

19.1 Implementing Blockchain in Cybersecurity

Since a Blockchain system is protected with the help of ledgers and cryptographic keys, attacking and manipulating it becomes extremely difficult. Blockchain decentralizes the systems by distributing ledger data on several systems rather than storing them on one single network. This allows the technology to focus on gathering data rather than worrying about any data being stolen. Thus, decentralization has led to an improved efficiency in Blockchain-operated systems.

For a Blockchain system to be penetrated, the attacker must intrude into every system on the network to manipulate the data that is stored on the network. The number of systems stored on every network can be in millions. Since domain editing rights are only given to those who require them, the hacker will not get the right to edit and manipulate the data even after hacking a million of systems. Since such manipulation of data on the network has never taken place on the Blockchain, it is not an easy task for any attacker.

While we store our data on a Blockchain system, the threat of a possible hack gets eliminated. Every time our data is stored or inserted into Blockchain ledgers, a new block is created. This block further stores a key that is cryptographically created. This key becomes the unlocking key for the next record that is to be stored onto the ledger. In this manner, the data is extremely secure.

Furthermore, the hashing feature of Blockchain technology [102] is one of its underlying qualities that makes it such a prominent technology. Using cryptography and the hashing algorithm, Blockchain technology converts the data stored in our ledgers. This hash encrypts the data and stores it in such a language that the data can only be decrypted using keys stored in the systems. Other than cybersecurity, Blockchain has many applications in several fields that help in maintaining and securing data. The fields where this technology is already showing its ability are finance, supply chain management and Blockchain-enabled smart contracts [103].

19.2 Advantages of Using Blockchain in Cybersecurity

The main advantages of Blockchain technology in cybersecurity (Figure 19.1) are the following [98, 99, 100, 120]:

19.2.1 Decentralization

Thanks to the peer-to-peer network, there is no need for third-party verification, as any user can see network transactions.

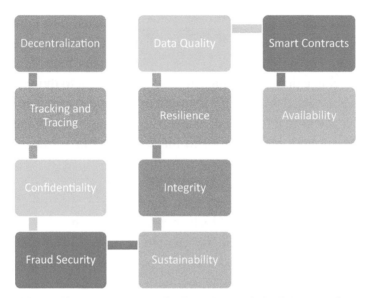

Figure 19.1   Advantages of using Blockchain in Cybersecurity.

19.2.2 Tracking and Tracing

All transactions in Blockchains are digitally signed and time-stamped, so network users can easily trace the history of transactions and track accounts at any historical moment. This feature also allows a company to have valid information about assets or product distribution.

19.2.3 Confidentiality

The confidentiality of network members is high due to the public-key cryptography that authenticates users and encrypts their transactions.

19.2.4 Fraud Security

In the event of a hack, it is easy to define malicious behavior due to the peer-to-peer connections and distributed consensus. As of today, Blockchains are considered technically "unhackable", as attackers can impact a network only by getting control of 51% of the network nodes.

19.2.5 Sustainability

Blockchain technology has no single point of failure, which means that even in the case of DDoS attacks, the system will operate as normal, thanks to multiple copies of the ledger.

19.2.6 Integrity

The distributed ledger ensures the protection of data against modification or destruction. Besides, the technology ensures the authenticity and irreversibility of completed transactions. Encrypted blocks contain immutable data that is resistant to hacking.

19.2.7 Resilience

The peer-to-peer nature of the technology ensures that the network will operate round-the-clock even if some nodes are offline or under attack. In the event of an attack, a company can make certain nodes redundant and operate as usual.

19.2.8 Data Quality

Blockchain technology cannot improve the quality of your data, but it can guarantee the accuracy and quality of data after it is encrypted in the Blockchain.

19.2.9 Smart Contracts

These are software programs that are based on the ledger. These programs ensure the execution of contract terms and verify parties. Blockchain technology can significantly increase the security standards for smart contracts, as it minimizes the risks of cyber-attacks and bugs.

19.2.10 Availability

There is no need to store your sensitive data in one place, as Blockchain technology allows you to have multiple copies of your data that are always available to network users.

19.2.11 Increase Customer Trust

Your clients will trust you more if you can ensure a high level of data security. Moreover, Blockchain technology allows you to provide your clients with information about your products and services instantly.

19.3 Disadvantages of Using Blockchain in Cybersecurity [98, 99, 100] (Figure 19.2)

19.3.1 Irreversibility

There is a risk that encrypted data may be unrecoverable in case a user loses or forgets the private key necessary to decrypt it.

19.3.2 Storage Limits

Each block can contain no more than 1 Mb of data, and a Blockchain can handle only seven transactions per second in average.

19.3.3 Risk of Cyberattacks

Although the technology greatly reduces the risk of malicious intervention, it is still not a panacea to all cyber-threats. If attackers manage to exploit the majority of your network, you may lose your entire database.

19.3.4 Adaptability Challenges

Although Blockchain technology can be applied to almost any business, companies may face difficulties integrating it. Blockchain

Figure 19.2   Disadvantages of using Blockchain in Cybersecurity.

applications can also require complete replacement of existing systems, so companies should consider this before implementing the Blockchain technology.

19.3.5 High Operation Costs

Running Blockchain technology requires substantial computing power, which may lead to high marginal costs in comparison to existing systems.

19.3.6 Blockchain Literacy

There are still not enough developers with experience in Blockchain technology and with deep knowledge of cryptography.

19.4 Conclusion

Blockchain's decentralized approach to cybersecurity can be seen as a fresh take on the issues that the industry faces today. The market could only use more solutions to combat the threats of cyberattacks. And the use of Blockchain may yet address the vulnerabilities and limitations of current security approaches and solutions.

Throwing constant pots of money at the problem and knee-jerk reactions is not the answer. Firms need to sort out their governance, awareness and organizational culture and critically look at the business purpose and processes before they invest in systems to combat cybercrime.

The roster of these new services provided by Blockchain may be limited for now and of course they face incumbent players in the cybersecurity space. But this only offers further opportunity for other ventures to cover other key areas of cybersecurity. Blockchain also transcends borders and nationalities, which should inspire trust in users. And, with the growth of these new solutions, the industry may yet restore some of the public's trust they may have lost in the midst of all these issues.

Overall, Blockchain technology is a breakthrough in cyber security, as it can ensure the highest level of data confidentiality, availability and security. However, the complexity of the technology may cause difficulties with development and real-world use.

Implementation of Blockchain applications requires comprehensive, enterprise- and risk-based approaches that capitalize on cyber-security risk frameworks, best practices and cybersecurity assurance services to mitigate risks. In addition, cyber intelligence capabilities, such as cognitive security, threat modeling and artificial intelligence, can help to proactively predict cyber threats to create counter measures. That is why, AI is considered as the first line of defense while Blockchain is the second line [112].

20

Future Trends of Blockchain

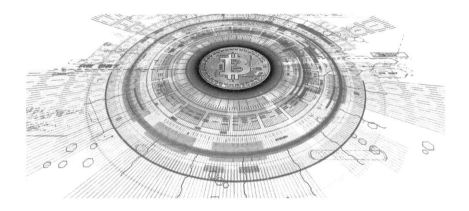

The year 2018 is the year of Blockchain applications with several ongoing use-cases coming to realization and the vendor landscape also gained more depth and a better structure after years of press and vendor hype, fueled equally by commercial self-interest and a genuine desire for innovation. Firms like Forrester predict that we will see more serious Blockchain projects in the coming years [119]. Initiatives will increasingly be assessed against standard business benefit models, and those found not up to standards will not be given the go-ahead, or they will be stopped if already underway.

20.1 Blockchain Tracks

To understand the future direction of Blockchain technology, we need to recognize the three tracks (Figure 20.1) of Blockchain technology [105]:

- *Pure R&D Track*: This track is focused on understanding what it means to develop a Blockchain-based system. Ideally, working on real use-cases, the ultimate goal is investigation and learning, and not necessarily delivery of a working system.
- *Immediate Business Benefit Track*: This track covers two bases: (1) learning how to work with this promising technology and (2) delivering an actual system that can be deployed in a real business context. Many of these projects are intra-company.

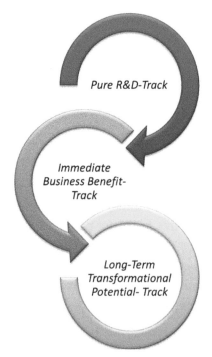

Figure 20.1   Tracks of Blockchain Technology.

- *Long-Term Transformational Potential Track*: This is the track of the visionaries, who recognize that to realize the true value of Blockchain-based networks means reinventing entire processes and industries as well as how public-sector organizations function.

20.2 Blockchain Technology Future Trends

Blockchain Technology future is bright and promising with the following trends shaping it (Figure 20.2):

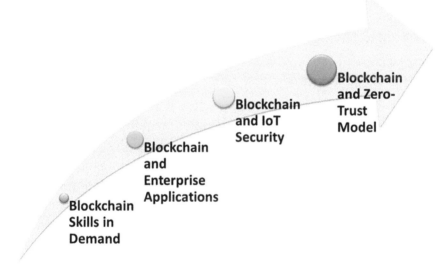

Figure 20.2   Blockchain Technology Future Trends.

20.3 Blockchain Skills in Demand

Opportunities for IT professionals with Blockchain skills are increasing at an astronomical rate. Blockchain is considered one of the most in-demand skills today, with Blockchain technologists taking second place in the top freelance skills list for the 2017's third quarter, according to Upwork. The numbers of job postings that feature Blockchain as a must-have skill have increased by 115% between 2016 and 2017 alone, according to Burning Glass Technologies, a job data analytics firm [110, 111].

The driving forces for this high demand are:

Cryptocurrencies: Part of why Blockchain has become so important in the business world is the rise of cryptocurrencies like Bitcoin. This makes the skill of particular interest to finance companies, such as banks and investment firms, as well as technology and consulting firms, and even application developers. With cryptocurrencies becoming a larger part of the market, large corporations are working to be able to accept or interact with currencies directly, increasing the need for Blockchain technologists.

Smart Contracts: Blockchain has introduced a new paradigm for recording information. This has to do with the distributed ledger systems created with Blockchain, an approach that increases efficiency and eliminates certain central administration requirements. Many industries are investigating the technology for its ability to support smart contracts, allowing specific actions to execute automatically as soon as the agreed upon conditions are met. Simple definition of a smart contract: A complex set of software codes with components designed to automate execution and settlement. It is the application layer necessary to make Blockchain technology a reality.

20.4 Blockchain and Enterprise Applications

Numerous major firms on Wall Street are looking to exponentially increase their experimentation with Blockchain technologies in 2018, meaning that trillions of dollars in institutional funds could be moving through the crypto space in the months ahead.

Powerhouse firms on Wall Street, like IBM, are helping companies create the infrastructure necessary for shifting operations to the Blockchain. Microsoft, too, has deemed cryptotech as enterprise-grade, and thus the global software giant is creating middleware suites that will make it more attractive and easier than ever for Wall Street to embrace Blockchain innovations.

Ongoing, reputable developments like this will surely instigate an avalanche of Wall Street Blockchain use-cases in 2018 and beyond, undoubtedly giving the Blockchain unprecedented prestige in the mainstream [106–110].

20.5 Blockchain and IoT Security

It is ironic that the cybercriminals who perpetrated the WannaCry ransomware attack in 2017 could hold a federal government to ransom and demand to be paid in Bitcoin [113]. Bitcoin might be a cryptocurrency, but it is based on Blockchain, and if cybercriminals are confident that Bitcoin provides a safe mechanism for the payment of ransoms, it indicates just how secure the distributed ledger approach is. Blockchain has the potential to totally re-engineer cybersecurity, but the industry has yet to come to terms with it.

Blockchain will deliver on the promise of Internet of Things (IoT) in the year ahead. In the world of IoT, you are generating millions of small transactions that are being collected from a distributed set of sensors. It is not feasible to operate these systems using a centralized transactional model: it is too slow, expensive and exclusive. To extract the true value from IoT technology, you have to be able to operate in real time. Once a sensor alert is received from a control system, you must react to it, meter it and bill for it instantly – all of which negates the viability of a centralized transactional authority. The cost of the transaction has to be near-zero or free, and the cost elements of a centralized model simply do not support the potential business model in IoT.

Recently some interesting applications of Blockchain and IoT in the area of cybersecurity emerged. Significant attacks have recently been launched from low-cost IoT endpoints, and there is very little incentive for manufacturers of these devices to incur the cost of a security stack, which leaves them extremely vulnerable. Blockchain can play a fundamental role in securing these environments.

20.6 Blockchain and Zero-Trust Model

Soon will see the "zero trust" security model re-emerging, which means that enterprise systems will vigorously authenticate whether users are indeed entitled access to specific sets of data, before making them available.

The Zero-Trust Model (ZTM) simplifies how information security is conceptualized by assuming that there are no longer "trusted" interfaces, applications, traffic, networks or users. It takes the old model – "trust but verify" – and inverts it because recent breaches have proven that when an organization trusts, it does not verify [104].

It requires that the following rules be followed:

- All resources must be accessed in a secure manner.
- Access control must be on a need-to-know basis and strictly enforced.
- Systems must verify and never trust.
- All traffic must be inspected, logged and reviewed.
- Systems must be designed from the inside out instead of the outside in.

So, essentially Blockchain will become the implementer of the "zero trust" policy.

20.7 Final Words

Those who persist with their Blockchain initiatives are not only aware, sometimes painfully, that the technology is still at a very early stage of development, but also understand that this is not really about technology, but about business. This is what sets them apart from those who follow the siren call of tech industry promises without sufficient grasp of what a Blockchain network is all about, both from a business and a technology perspective; the resulting vanity projects will invariably fail [109].

References

[1] https://www.goldmansachs.com/insights/pages/internet-of-things/

[2] https://openconnectivity.org/developer/reference-implementation/alljoyn

[3] https://www.cio.com/article/2872574/it-industry/5-key-challenges-facing-the-industrial-internet-of-things.html?page=2

[4] https://www.recode.net/2015/1/22/11557992/reimagining-business-with-the-industrial-internet-of-things

[5] https://www.accenture.com/us-en/new-applied-now

[6] https://www.zdnet.com/article/welcome-to-the-dystopian-internet-of-things-powered-by-and-starring-you/

[7] https://www.bbvaopenmind.com/en/the-internet-of-everything-ioe/

[8] http://www.eweek.com/blogs/first-read/cisco-ceo-internet-of-everything-will-be-worth-19-trillion

[9] https://www.idc.com/getdoc.jsp?containerId=prUS43994118

[10] http://www.eweek.com/security/cisco-sponsors-300-000-internet-of-things-security-challenge

[11] https://www.ndtv.com/business/internet will-disappear-google-boss-tells-davos-731773

[12] https://www.csoonline.com/article/2134066/mobile-security/what-the-internet-of-things-means-for-security.html

[13] https://www.sans.org/instructors/paul-a-henry

[14] https://security.vt.edu/about/faculty_staff/randy_marchany.html

[15] https://www.owasp.org/index.php/Main_Page

[16] http://www.woodsidecap.com/wp-content/uploads/2015/03/WCP-IOT-M_and_A-REPORT-2015-3.pdf

[17] https://www.bbvaopenmind.com/en/internet-of-things-iot-the-third-wave/

[18] https://www.idc.com/getdoc.jsp?containerId=prUS25291514

[19] https://www.linkedin.com/pulse/industrial-internet-things-iiot-challenges-benefits-ahmed-banafa/?trk=mp-reader-card

[20] https://www.bbvaopenmind.com/en/iot-securityprivacy-and-safety/

[21] https://www.pushtechnology.com/hs-fs/hub/188121/file-1221572117-pdf/Myths_of_IoT_white_paper_FINAL.pdf

[22] https://www.cisco.com/c/en/us/solutions/collateral/enterprise/cisco-on-cisco/Cisco_IT_Trends_IoE_Is_the_New_Economy.html

[23] https://www.cisco.com/c/en/us/solutions/internet-of-things/smart-city-infrastructure-guide.html#~stickynav=1

[24] https://www.linkedin.com/pulse/20140520183722-246665791-fog-computing/?trk=mp-reader-card

[25] http://www.mouser.com/pdfdocs/IOTEXECBRIEFWP.PDF

[26] https://www.linkedin.com/pulse/20140312180810-246665791-the-future-of-big-data-and-analytics/?trk=mp-reader-card

[27] http://www.microwavejournal.com/articles/27690-addressing-the-challenges-facing-iot-adoption

[28] https://blog.apnic.net/2015/10/20/5-challenges-of-the-internet-of-things/

[29] https://www.sitepoint.com/4-major-technical-challenges-facing-iot-developers/

[30] https://www.linkedin.com/pulse/iot-implementation-challenges-ahmed-banafa?trk=mp-author-card

[31] https://www.bbvaopenmind.com/en/why-iot-needs-fog-computing/

[32] http://iot.ieee.org/newsletter/january-2017/iot-and-blockchain-convergence-benefits-and-challenges.html

[33] https://ieeexplore.ieee.org/document/5739775/

[34] https://www2.deloitte.com/insights/us/en.html

[35] http://www.dbta.com/BigDataQuarterly/Articles/10-Predictions-for-the-Future-of-IoT-109996.aspx

[36] https://campustechnology.com/articles/2016/02/25/security-tops-list-of-trends-that-will-impact-the-internet-of-things.aspx

[37] http://dupress.com/

[38] https://datafloq.com/read/iot-standardization-and-implementation-challenges/2164

[39] https://www.bbvaopenmind.com/en/what-is-next-for-iot/

[40] https://www.sap.com/documents/2017/06/e825c3a3-c27c-0010-82c7-eda71af511fa.html

[41] https://www.bbvaopenmind.com/en/understanding-dark-data/

[42] https://www.idc.com/infographics/IoT

[43] https://www.bbvaopenmind.com/en/data-lake-an-opportunity-or-a-dream-for-big-data/

[44] https://datafloq.com/read/why-ai-is-the-catalyst-of-iot/3046

[45] https://www.bbvaopenmind.com/en/cloud-computing-big-data-and-mobility-2015-tech-trends/

[46] https://datafloq.com/read/fog-computing-vital-successful-internet-of-things/1166

[47] https://www.idc.com/getdoc.jsp?containerId=prUS25291514

[48] https://cloudsecurityalliance.org/

[49] https://www.gartner.com/newsroom/id/3123018

[50] https://techcrunch.com/2016/06/28/decentralizing-iot-networks-through-blockchain/

[51] http://www-935.ibm.com/services/multimedia/GBE03662USEN.pdf

[52] https://www.computerworld.com/article/3027522/internet-of-things/beyond-bitcoin-can-the-blockchain-power-industrial-iot.html

[53] http://www.treasuryandrisk.com/2017/03/09/blockchain-technology-balancing-benefits-and-evolv?slreturn=1507334668&page=4

[54] http://www.livebitcoinnews.com/three-risks-assess-company-considering-blockchain/

[55] https://hbr.org/2017/03/how-safe-are-blockchains-it-depends

[56] http://tech.economictimes.indiatimes.com/news/internet/5-challenges-to-internet-of-things/52700940

[57] http://www.gartner.com/newsroom/id/3221818

[58] http://www.gartner.com/smarterwithgartner/top-10-security-predictions-2016/

[59] http://www.mindanalytics.es/2016/03/01/gartners-top-10-internet-of-things-technologies-for-2017-2018/?lang=en

[60] http://www.cnbc.com/2016/10/22/ddos-attack-sophisticated-highly-distributed-involved-millions-of-ip-addresses-dyn.html

[61] https://krebsonsecurity.com/2016/10/hacked-cameras-dvrs-powered-todays-massive-internet-outage/

[62] https://www.cisco.com/c/dam/en_us/solutions/trends/iot/docs/computing-overview.pdf

[63] https://www.datacenterknowledge.com/archives/2015/04/08/fog-computing-for-internet-of-things-needs-smarter-gateways

[64] https://biztechmagazine.com/article/2014/08/fog-computing-keeps-data-right-where-internet-things-needs-it

[65] https://aibusiness.com/ai-brain-iot-body/

[66] http://www.creativevirtual.com/artificial-intelligence-the-internet-of-things-and-business-disruption/

[67] https://www.computer.org/web/sensing-iot/contentg=53926943&type=article&urlTitle=what-are-the-components-of-iot-

[68] https://www.bbvaopenmind.com/en/the-last-mile-of-iot-artificial-intelligence-ai/

[69] http://www.datawatch.com/

[70] https://www.pwc.es/es/publicaciones/digital/pwc-ai-and-iot.pdf

[71] http://www.iamwire.com/2017/01/iot-ai/148265

[72] https://www.i-scoop.eu/industry-4-0/

[73] https://www.technologyreview.com/s/603298/a-secure-model-of-iot-with-blockchain/

[74] http://www.zdnet.com/article/why-ai-and-machine-learning-need-to-be-part-of-your-digital-transformation-plans/

[75] https://www.i-scoop.eu/digital-transformation/

[76] https://www.linkedin.com/pulse/iot-blockchain-convergence-ahmed-banafa/

[77] https://iot.ieee.org/newsletter/january-2017/iot-and-blockchain-convergence-benefits-and-challenges.html

[78] http://www.ioti.com/strategy/five-internet-things-trends-watch

[79] https://mobidev.biz/blog/iot-trends-for-business-2018-and-beyond

[80] https://www.bayshorenetworks.com/blog/breaking-down-idc-top-10-iot-predictions-for-2017

[81] https://readwrite.com/2017/10/03/6-iot-trends-2018/

[82] https://lightingarena.com/internet-things-anticipated-trends-2018/

[83] https://medium.com/@Unfoldlabs/seven-trends-in-iot-that-will-define-2018-2a47e763731c

[84] https://datafloq.com/read/iot-and-blockchain-challenges-and-risks/3797

[85] https://www.bbvaopenmind.com/en/five-challenges-to-iot-analytics-success/

[86] https://www.bbvaopenmind.com/en/why-iot-needs-ai/

[87] https://www.bbvaopenmind.com/en/a-secure-model-of-iot-with-blockchain/

[88] https://datafloq.com/read/fog-computing-vital-successful-internet-of-things/1166

[89] https://iot.ieee.org/newsletter/july-2016/iot-standardization-and-implementation-challenges.html

[90] https://ripple.com/insights/busting-myths-around-blockchains/

[91] http://www.reuters.com/article/us-blockchains-technology-commentary-idUSKCN0Y22GC

[92] https://www.forbes.com/sites/yec/2017/05/04/debunking-blockchain-myths-and-how-they-will-impact-the-future-of-business/#275f5ef05609

[93] http://www.techrepublic.com/article/five-big-myths-about-the-bitcoin-blockchain/

[94] https://home.kpmg.com/uk/en/home/insights/2017/04/five-blockchain-myths-that-just-wont-die.html

[95] https://www.allerin.com/blog/4-myths-associated-with-blockchain

[96] https://blockchainhub.net/blockchains-and-distributed-ledger-technologies-in-general/

[97] https://datafloq.com/read/how-blockchain-secure-model-internet-of-things/2583

[98] https://www.ibm.com/blogs/insights-on-business/government/convergence-blockchain-cybersecurity/

[99] https://www.forbes.com/sites/rogeraitken/2017/11/13/new-blockchain-platforms-emerge-to-fight-cybercrime-secure-the-future/#25bdc5468adc

[100] http://www.technologyrecord.com/Article/cybersecurity-via-blockchain-the-pros-and-cons-62035

[101] https://www.allerin.com/blog/blockchain-cybersecurity

[102] https://www.infosecurity-magazine.com/next-gen-infosec/blockchain-technology/

[103] https://www.allerin.com/blog/blockchain-enabled-smart-contracts-all-you-need-to-know

[104] https://www.computerworld.com/article/2851517/network-security-needs-big-data.html

[105] http://www.zdnet.com/article/how-blockchain-will-shape-up-in-the-enterprise-in-2018/

[106] https://www.finextra.com/blogposting/14151/blockchain-technology-by-2018-a-breakthrough

[107] https://esg-intelligence.com/blockchain-articles/2017/11/13/blockchain-challenges-2018/

[108] https://www.baysidesolutions.com/2017/11/13/blockchain-technologists-will-high-demand-2018/

[109] http://www.cryptoanalyst.co/2017/10/02/2018-wall-street-blockchain-trillions/

[110] https://cio.economictimes.indiatimes.com/news/strategy-and-management/blockchain-machine-learning-robotics-artificial-

intelligence-and-wireless-technologies-will-reshape-digital-business-in-2018/61198007

[111] https://www.accountingtoday.com/news/ey-to-hire-over-14-000-candidates-in-2018

[112] https://www.upwork.com/press/2018/05/01/q1-2018-skills-index/

[113] https://www.bbntimes.com/en/technology/first-line-of-defense-for-cybersecurity-artificial-intelligence

[114] https://www.cnet.com/news/wannacry-wannacrypt-uiwix-ransomware-everything-you-need-to-know/

Index

A

actuators 8, 108, 117
Amazon 108
Apple 4, 108
Artificial Intelligence
(AI) 63, 107
ATM 121
authentication 17, 75, 91, 96
Autonomous 15, 26, 62, 84

B

Big Data 5, 39, 110, 126
Bitcoin 80, 133, 149
Blockchain Certified 124
Blockchain Literacy 143
Blockchain 38, 82, 123, 150
Blockchain-as-a-
Service 90, 133
Blocks 25, 81, 118, 141
Bluetooth 47, 56
Business 11, 50, 112, 150
Business model 11, 55,
126, 149
BYOD 17, 24

C

Capex 126

CEP 39, 50
Cloud infrastructure 70, 79
cognitive technologies 39,
49, 57
Communications 9, 29, 71, 95
Computational Logic 87
Computer vision 50
Confidentiality 132, 140, 144
Connectivity 4, 54, 95, 125
consensus 80, 124, 131, 140
Consortium 29
Corda 133
Credential Security 90
Cryptocurrency 123, 135, 149
CSA 75
Customer Trust 142
Cyberattacks 143
Cybersecurity 14, 94,
137, 149

D

Dark Data 61, 64, 110
Data 4, 23, 87, 150
Data Analysis 64, 110, 126
Data Discovery 64, 110, 125
Data Lakes 64, 110, 132
Data Quality 141

Data Structures 61
Datacenters 26, 28
DDoS 16, 141
Deep Learning 40, 51, 68
Digital Transformation
    115, 118
Distributed Ledger 80,
    132, 149
DX 116

E
Ethereum 133
ETL 48

F
Federated Blockchain 131
Fog Computing 26, 62,
    105, 125
Forrester 145
Fourth Industrial
    Revolution 108
Fraud 63, 132, 140

G
Gartner 74, 95, 105, 134
Gateways 27, 70, 105
Gbps 62
Google 15, 108

H
Hacked 16, 134
Hacking 16, 139, 141
hash 82, 86
hashing 139

HIPAA 49
HTTP 48
Hyperledger 133

I
IaaS 48, 57
IBM 133, 148
IDC 14, 102, 124
IEEE 29, 46, 122
Industrial Internet of Things
    (IIoT) 8, 12, 55
Industry 4.0 108
Integrity 64, 110, 134, 141
Intelligent Actions 40, 51, 58
Intelligent Analysis 36, 46, 57
Internet of Things (IoT) 3, 78,
    126, 149
IoT Value Chain 23
IoT-as-a-Service
    (IoT-a-a-S) 125
IP 5, 22, 69
IPv6 48
Irreversibility 87, 141
ISA 29
ISOC 95

K
Killer applications 54, 122

L
LAN 47
Legacy systems 39, 50, 58
Legal and Compliance 91
life cycle 71

Li-Fi 47, 56
logistical Data 64, 110, 125
LTE 27, 47, 56

M
M2H 40, 51, 58
M2M 9, 38, 58, 79
MAN 47
Microsoft 133, 148
Mirai 94

N
Natural-language
   processing 50
NetFlix 94
Network infrastructure 70
Networks 15, 74, 134, 150
NoSQL 39, 49
NYTimes 94

O
OEM 97
Opex 126
Optimum Platform 74, 78
OWASP 18

P
PaaS 48, 57
PayPal 94
Peer-to-peer 38, 83, 135, 141
Predictive and Advance
   Analytics 64, 125
Privacy 6, 42, 80, 123

Private Blockchain 82,
   88, 134
Processing power 4, 21,
   88, 134
PSS 111
Public Blockchain 82, 88, 138

R
Real-Time Geospatial and
   Location 64, 110, 125
Redundancy 105
Resilience 96, 105, 141

S
SaaS 48, 57
Security 9, 78, 150
self-driving cars 109
Sensors 4, 55, 124, 149
single version of truth
   (SVT) 133
Siri 108
Smart Cars 41, 54
Smart Cities 8, 24, 55, 126
Smart Contracts 84, 136,
   141, 148
Smart Devices 5, 38, 89, 123
Society 8, 35, 95, 116
Speech recognition 50
SQL 39, 49
Standardization 58, 122
Storage 39, 86, 104, 142
Streaming Data 64, 110, 125
Sustainability 141

T
terabyte 62, 109
Time Series Accuracy
    of Data 64, 110, 125
Transactions 73, 118, 149
Transparency 43, 87, 134, 138
Twitter 94

U
UI 40, 58
UX 9, 40, 58

V
Vendor Risks 90
verification 18, 88, 91, 139
VUI 123

W
W3C 29
WAN 47, 97
Wi-Fi 4, 27, 56, 123
Wi-Max 47, 56
wireless communications 71

Z
Zero-Trust Model 150
(ZTM)

About the Author

Prof. Ahmed Banafa has extensive experience in research, operations and management, with focus on IoT, Blockchain, Cybersecurity and AI. He is a reviewer and a technical contributor for the publication of several technical books. He served as an instructor at well-known universities and colleges, including the Stanford University, University of California, Berkeley; California State University-East Bay; San Jose State University; and University of Massachusetts. He is the recipient of several awards, including Distinguished Tenured Staff Award, Instructor of the year for 4 years in a row, and Certificate of Honor from the City and County of San Francisco. He was named as No.1 tech voice to follow, technology fortune teller and influencer by LinkedIn in 2018 by LinkedIn, his researches featured in many reputable sites and magazines including Forbes, IEEE and MIT Technology Review, and Interviewed by ABC, CBS, NBC and Fox TV and Radio stations. He studied Electrical Engineering at Lehigh University, Cybersecurity at Harvard University and Digital Transformation at Massachusetts Institute of Technology (MIT).

Milton Keynes UK
Ingram Content Group UK Ltd.
UKHW051536141024
449569UK00001B/53